Agricultural Economics

Dr. Pranav K. Desai is having more than seventeen years of teaching and research experience, mainly on Agricultural Finance. Throughout his career he has been associated with a number of educational and research institutes. He has published a number of research articles in reputed journals. His areas of interest includes agricultural economics and natural resource management.

Apart from being an active and dedicated teacher, he has also been committed in research and running projects funded by the eminent bodies.

Agricultural Economics

by

Dr. Pranav K. Desai

BIOTECH

2023

Biotech Books

ISBN 81-7622-208-9
ISBN 978-81-7622-208-2

Published by : **BIOTECH BOOKS**
7462-63/23, Ansari Road, Darya Ganj,
New Delhi - 110002
Phones : 011 4300 3222
Email : biotechbooks@yahoo.co.in

PRINTED IN INDIA

Preface

Agricultural Economics generally applies the principles of economics to the production of crops and livestock. The discipline, known as Agronomics, is a branch of Economics, which specifically deals with land usage. It focuses on maximising the yield of crops, while maintaining a good soil-ecosystem. Throughout the 20th century, the discipline got expanded and the current scope of the discipline is much broader. Agricultural Economics now includes a variety of applied areas, having considerable overlap, with conventional economics.

Economics is the study of resource allocation under scarcity. Agronomics or the application of economic methods to optimising the decisions, made by agricultural producers, grew to prominence, around the turn of 20th century. The field of agricultural economics can be traced out to works on land economics. The field of agricultural economics has evolved over many decades.

Agricultural economists have made many a well-known contribution to the field of economics, with such models as the cobweb model, hedonic regression, pricing models, new technology, diffusion models (Zvi Griliches), multifactor productivity and efficiency theory and measurement, and the random coefficient regression. The farm sector is frequently cited as a prime example of the perfect competition in economic paradigm.

Since 1970s, agricultural economics has primarily focused on seven main topics; technical change and human capital, agricultural environment and resources, risk and uncertainty, consumption and food supply chains, prices and income, market structure and trade and development. Agricultural economics tends to be more microeconomics oriented.

There is a dearth of books on Agricultural Economics. Hence, this effort. Expectedly, this research-based, compact and exclusive study would fill the room for a reader-friendly reference book on the subject.

– *Pranav K. Desai*

Contents

1

Introduction

It is only recently that agricultural economics has come to be studied as a separate branch of economics and agronomy. Although books on agriculture had been written in the past and many important problems of agricultural economics given considerable thought and solved through organised efforts, not much of literature exists on agricultural economics as a separate discipline. Agricultural economics as a separate discipline started only in the beginning of 20th century when interest in economic issues related to agriculture erupted in several educational centres. It was only after the depression of 1890's which seriously affected agriculture, that organised farm groups stirred considerable interest in farm management problems. This new field of agricultural interest was later designated as agricultural economics.

It is easy to understand that the whole of economics in the past must have been written, taught and explained through illustrations drawn from agriculture either directly or indirectly. Not before the industrial revolution could one imagine of understanding economics without reference to agriculture. The emergence of industrial revolution did not mar the interest and development of agriculture rather stimulated its growth and facilitated understanding of its importance. This stimulation and understanding slowly and gradually resulted in the evolution of agricultural economics were given by agronomists. However, the importance of this discipline has grown to much that we have now separate agricultural universities and agricultural economics as an inseparable part of the academic programmes.

The courses on agricultural economics are given in other academic institutions also and as different levels of learning with of course, not much emphasis on theory. Applied aspects in relation to local conditions are given weightage in teaching programmes in these institutions.

Concept of Agricultural Economics

The words agricultural economics are made up of two words, *viz.*, agriculture and economics. Before we define agricultural economics as a whole, it would be appropriate to define agriculture and economics separately.

The word agriculture, since long, has been associated with the industry of basic food production, known as farming. Agriculture and farming were synonymous till farming was not commercialised and was done more or less on a subsistence basis. But after the commercialisation of the agricultural sector, production of food and fibre (farming) has become only one part of scientific agriculture.

Modern agriculture has much wider scope today and it includes the farm supply industries as well as the product-processing industries and distribution industries, too. Industries closely related to farming are known as agriculturally related industries or agribusinesses. The scope of modern agriculture has been depicted in the following Fig.

Agriculture thus, may be defined as the production, processing, marketing and distribution of crops and livestock. These four activities were previously all farm-centred. However, with improvement in technology, transport and communication developments and specialisation of labour, some of these activities have moved away from farm into certain strategic control points. Modern agriculture also includes the farm supporting industries as well as product processing and distribution industries.

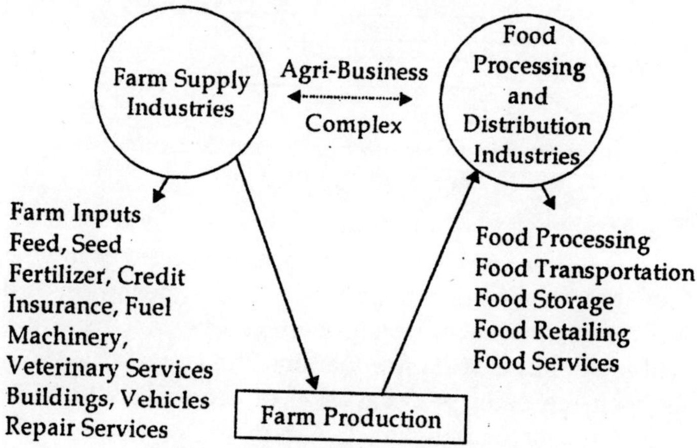

Fig: Scope of Modern Agriculture.

Economics, on the other hand, is the science of analysing the use of limited resources to achieve desired ends. It is a social science which studies how man satisfies wants through the allocation of scarce resources.

The English word "economics" is derived from the Greek word "Oekonomous" which means "housekeeping" and one of the tasks of the housewife is of making most effective use of the resources that are available to her, Prof. Lionel Robbins has defined economics as "the science which studies human behaviour as a relationship between ends and scarce means which have alternative uses". As such it involves.

1. Deciding between alternative ways of using limited resources;
2. Satisfying human needs and wants for which there are varying degrees of preferences; and
3. Taking into account human behaviour and decision-making on the best way to use available resources.

Fig. illustrates the general subject-matter underlying the study of economics:

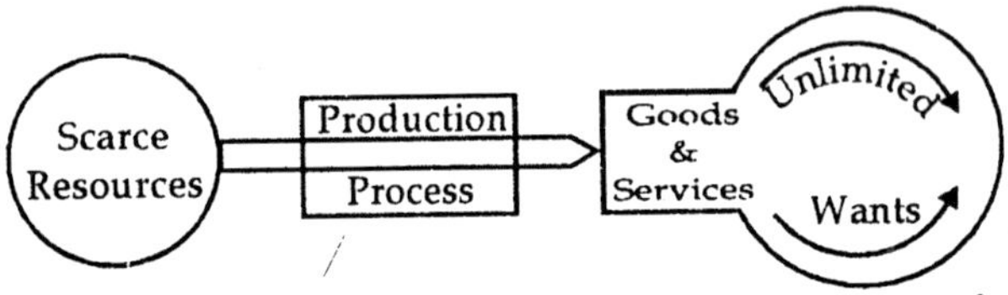

Fig: Scarce Resources Attempt to Satisfy Unlimited Wanted via Production.

The gap between what people wish to have and what they actually can afford to have is of central concern in economics. In the absence of such a gap between expectations and achievements, the subject of economics may not have been studied at all. But in real life situations, the gap does exist and it has been the main preoccupation of economists to evolve ways and means of reducing this gap.

Nature of Agricultural Economics

After having discussed the definitions of agriculture and economics separately, we are now in a position to define agricultural economics.

In a very simple language, agricultural economics may be defined as an applied phase of economics in which attention is given to all aspects of problems related to agriculture. It helps the farmer in deciding about what kind of food should he produce, which crop should he raise to maximise his profits and at what level should he price his products. As such, it is a social science concerned with the allocation of scarce resources among the uses associated with producing, processing and consuming the farm products. According to Prof. Gray, "agriculture economics may be defined as the science in which the principles and methods of economics are applied to the special conditions of agricultural industry."

According to A.W. Ashby, "agricultural economics is an applied science, that is, it is a methodical pursuit of knowledge of economic process and organisation in agriculture and of their results, for the purpose of stabilising, adopting or modifying, and if and when necessary, of changing their results."

Hibbard defines agricultural economics as "a study of relationships arising from the wealth getting and wealth using activity of man in agriculture." However, agricultural economics should include within its purview not only subjects directly connected with the exploitation of land but also those which indirectly influence the economic activity on the farm and the well-being of the farm population.

To Jouzier, "agricultural economics is that branch of agricultural science which treats of the manner of regulating the relations of the different elements comprising the resources of the farmer, whether it be the relations to each other or to human beings, in order to secure the greatest degree of prosperity to the enterprise."

In short, we may define agricultural economics as an applied science which is mainly concerned with economic problems that are associated with farmers' effort to make a living. Their problems as we know are numerous and varied in character but can be grouped under three main heads: (i) Production; (ii) Marketing; and (iii) Financing. Thus, agricultural economics is concerned with the evolving of appropriate principles that govern the amount of land, labour and capital, which a farmer should use to farm most efficiently. Agricultural economics is equally concerned with the forces that affect the prices of the things he buys, sells or in other words, the relation between agriculture and the rest of the economy.

As an applied science, an important role of agricultural economists is to formulate the methods, techniques and procedures by which the problems of agriculture may be tackled. This perhaps is the most difficult function which the agricultural economist is called upon to perform. Hence we may conclude by saying that "agricultural economics is an applied science and as such is concerned with the identification, description and classification of the economic problems of agriculture, to the end that these problems may be solved."

The problem of allocation of scarce means to diverse uses is perhaps more crucial in the field of agriculture than in the economy as a whole for the simple reason that land – the basis of all agricultural pursuits – is highly limited in supply. The theoretical frame of agricultural economics has, therefore, to be more rigorous and thoughtful so as to provide a plausible kit of methods and procedures by which this most precious and limited resource is allocated among different productive activities to the maximum satisfaction of the society. Agricultural economics is, therefore, both theoretical and applied in its character. The theory of agricultural economics deals with the development of principles of resource economics and as an applied science, it deals with the application of these principles in diverse productive activities related to agriculture.

Scope of Agricultural Economics

There are a number of forces active in agriculture. Some of these forces are physical and are taken care of by physicians, chemists, geologists and soil scientists and so on. Some others are biological forces which are studied by plant and animal physiologists, pathologists, geneticists, entomologists, bacteriologists, etc. Agriculture is also faced with economic and sociological forces, which are characterised by relationships arising amongst men on account of agriculture being followed as a vocation and a way of life. It is with these relationships that agricultural economists deal. These relationships comprise of:

1. Relationship of contact (supplementary, complementary or competitive) between the different branches of the enterprise such as the simultaneous raising of cereals and animals on the same farm;
2. Relationship of activity between the different means employed in the process of production, as in the simultaneous employment of machinery and human labour;
3. Relations of value between the means employed in production and the product itself; and

4. Commercial relations with the people to whom farmers sell their products or from whom they buy their requirements.

> "The domain of agricultural economics then covers the examination of each element of agricultural production whether in connection with any one of the above-named relationships in particular or with several of them together, for the purpose of obtaining the greatest net profit."

It is clear from the above statement of Jouzier that he looks upon agricultural economics as *(i)* the science which deals with the principles underlying the coordination of all the factors involved in farming (land, labour, equipment and the various lines of production) in such a manner as will enable the farmer to secure the maximum net profits; and *(ii)* the art of applying these principles on a given farm.

We can thus describe the scope of agricultural economics in the words of Taylor who says, "agricultural economics deals with the principles which underline the farmers' problem of what to produce and how to produce it, what to sell and how to sell it in order to secure the largest net profit for himself consistent with the best interest of society as a whole. More specifically, it treats of the selection of land, labour and equipment for a farm, the choice of crops to be grown, the selection of livestock, enterprises to be carried on, and the whole question of the proportions in which all these agencies should be combined. These questions are treated primarily from the point of view of costs and prices."

Can we include in the scope of agricultural economics the distribution and processing of farm products after they leave the farm and the consumption of food and clothing by the urban population? The answer naturally would be "No"; otherwise, steel manufacturing and automobile industries would be treated as part of mining. But this does not mean that agricultural economics has nothing to do with the problems of distribution and consumption of farm products. Indeed, it is as important for farm people to understand the economics of distribution and of consumption of farm products as it is for them to understand the economics of their farm production. They are, therefore, part of the economics of agriculture."

The scope of agricultural economics is, thus, extended to the distribution and consumption problems of farm products as well; what to distribute, among whom to distribute and on what basis to distribute; what to consume and how much to consume are, therefore, some of the important questions dealt in agricultural economics. A study of agricultural economics also includes the functioning of government in agriculture. "To neglect the public management, public aid, and public regulation as arts of the agricultural economics would be almost like leaving mechanics out of a course of physics."

To be specific, the scope of agricultural economics can be stated to include the choice of farming as an occupation, the choice of farm and livestock, of machinery and labour, of crops and cropping system, the size of the unit of production, the grouping of the factors of production, intensity of cultivation, manuring, irrigation, soil conservation, selling of agricultural products, land system and rent, agricultural finance and rate of interest, wages and employment, prices, costs and profits, standard of living, national dividend, etc."

The main problem in all the cases mentioned above is of choosing the most profitable enterprise and operating it. The task of agricultural economist is to point out what is best to do in the economic interest of the agricultural community under the given conditions.

From the above, we may conclude that the scope of agricultural economics is very vast. "Within its scope must be included every phase of a nation's activities that in any way affect the farmer in his efforts to make a living. At the farm level, agricultural economics is concerned with developing ways and means by which each productive unit may be so organised and operated, its products so disposed of and the resources so allocated to various uses that the farm may yield maximum net returns. At the sector level, it deals with those problems that emerge when farmers attempt to effect changes in farming that cannot be accomplished by individual action. Among such problems are the conservation of land, the cooperative purchase of goods and services and the marketing of farm products.

Finally, agricultural economics must be concerned with solutions of those problems that require the active participation of the Government, *i.e.*, problems connected with price control, supply of credit, international trade, etc."

Types of Agricultural Economics

Agricultural economics is both a social science as also a natural science. As a social science, it does not differ significantly in its scope from general economics. A farmer's activities, decisions and functioning are greatly affected by society, which in turn, is equally affected by the farmer's decisions. To the extent farmer's decisions are affected by his own psychology, social institutions and religious and other taboos, his activities cannot be measured in as precise a manner as would be possible in a laboratory experiment. In this respect, agricultural economics is subject to all such influences as exist within the domain of social sciences. Economic calculations would, therefore, follow the same standards of accuracy or inaccuracy as in general economics.

On the other hand, agricultural economics as a natural science deals with a thorough examination and evaluation of scientific innovations suggested by soil scientists. In fact, agricultural economics owes its origin in the present form to agronomy itself.

Agricultural economics is essentially considered to be an applied science for the sole reason that it is the practical wisdom of an agricultural economist rather than the volume of his theoretical knowledge without practical orientation that makes the difference in agricultural pursuits. Knowledge is not to be acquired for the sake of knowledge alone but the theoretical and practical aspects have to be studied in relation to each other. By nature, the learning process in agriculture has to be such as would offer itself convincingly for use for the directive purpose of economic pursuits.

The nature of agricultural economics is unique because of the important role that nature plays in its working. In no other economic phenomenon as in agriculture is the nature so directly and strongly involved. The problem of maximisation of returns is, therefore, more complex and solution to the problem more uncertain. Varying agroclimatic conditions, economic systems, soil fertility and soil capacities lead to inter-regional and intra-regional heterogeneity in the agricultural production conditions.

This creates the problem of differential achievements under different geoeconomic conditions. The problems of agricultural production are, therefore, multidimensional and the subject of agricultural economics has to develop in the light of these problems that it seeks to investigate and work with.

Seasonality is yet another factor that distinguishes agricultural production enterprise from other non-agricultural activities. Naturally, therefore, market fluctuations are more sharp and frequent in agriculture. Since the very nature of agricultural enterprise is such that it involves the whole family in its pursuit for livelihood. It differs uniquely in its organisation and institutional setup from non-agricultural pursuits.

Relationship with other Disciplines

It would be a grave error to infer that economics is the only science upon which the agricultural economist depends. The applied or technical agricultural sciences such as agronomy, animal husbandry and agricultural engineering contribute data showing the effect of various types of practices and equipment on output. Such information is helpful or rather indispensable to the agricultural economist in dealing with problems of production efficiency.

Again, the determination of the effect of any technical practice requires that records and complete farm accounts be maintained. For assistance in this sphere, the agricultural economist depends upon the science of bookkeeping and accounting which has developed rules and procedures by which accurate data on any phase of the farm business can be obtained.

Statistics is another science that has been used effectively by the agricultural economists. This science is helpful in supplying methods by which data regarding specific farm problems can be collected, analysed and evaluated.

Agricultural economists must also draw upon the subject method of other social sciences. The process of decision-making under uncertainties involves psychology as such as it involves economics. Sociology is necessarily involved in production and resource-use. Interfarm problems in production are partly those of community organisation and the acceptance of production policies or new techniques of production involve sociological as well as economic considerations. Political science, if it is considered from the stand point of choice and decision, likewise becomes a field of science which must be integrated with the principles of production and consumption.

Physical resources and physical sciences specify what can be produced; economics specifies how resources should be used in production while sociology, psychology, ethics and political. Science specify the limitations which are placed on choice through laws, customs and other expression of individual and group values.

Industrial Development and Agriculture

If we study the economic history of most of the advanced countries, we find that it was revolution in the agricultural sector that gave rise to revolution in their industrial sector. Industry for its development needs capital for investment, labour for running the factories, raw materials as their inputs and demand for their products. A developed agriculture as we have seen above can contribute all these ingredients for the rapid development of the industrial sector of an economy. The developed industrial sector itself becomes an engine for rapid growth of agricultural sector.

Hence with the passage of time, agricultural sector begins to depend upon the industrial sector for its own development. In the later stages both sectors become interdependent. Today, it is futile to discuss the issue of industrial development versus agricultural development but the current thinking may be devoted to consider how best agriculture and industry contribute to each other.

Mostly there are two channels through which the sectors are linked to each other, *viz.*, through production and demand. Firstly, each requires inputs from the other sector. Agriculture requires industrial inputs such as fertilizer and pesticides. The agro-based industries require agricultural raw material as their inputs. Secondly, the incomes arising out of agricultural sector create demand for goods of industrial sector and the incomes of the industrial sector generate demand for agricultural commodities.

To quantify these linkages, the transactions that take place between the producing sectors are needed. The input-output tables precisely provide such a framework.

Significance of Agricultural Economics

Agricultural economics is not basically a different kind of economics – a separate set of economic principles and methodology which has relevance only to agriculture. The general framework of economic theory is applicable to the business of agriculture just as much as to that of industry. The analysis of equilibrium of demand and supply, of value and price, etc., is as valid in agriculture as in industry. Then if general economic principles are applicable to the agricultural sector the question arises: why should we study agricultural economics separately?

True, goals of production and the need for management decisions concerning the allocation of inputs are strikingly similar between agricultural and industrial production. However, there are substantial differences in the natural conditions under which production must be carried on and in the sociological background in the agricultural sector which demands a separate study of agricultural economics.

Firstly, agriculture is a unique industry in which the mode of life and business enterprise are combined together. This combination no longer exists in present-day industry. It is on this score that it is more influenced by sociological, political and sentimental considerations.

Secondly, another characteristic of agricultural production which distinguishes it from industrial production is that the farmer produces chiefly for his own needs. No denying the fact that in modern farming, the element of self-sufficiency has been reduced in importance but by no means has it been removed. In most underdeveloped countries, even today, farming is done on the basis of self-sufficiency.

Thirdly, many agricultural commodities are joint products like wheat and affals or mutton and wool because they are both part of the same plant or the same animal. The costs attributable to the various products cannot be separated as they often can be in industry even when several products are produced in the same plant. Thus, in agriculture, it is rarely justifiable to consider the supply of any product in isolation.

Fourthly, agriculture requires a far larger proportion of land in relation to its employment of other factors than does industry. This is responsible for an early tendency of law of diminishing returns wide scatter of production and the great importance of system of land tenure.

Fifthly, farming mostly is undertaken in small-sized units and thereby gives little scope for division of labour. Thus, large-scale organisation and its benefits, typical of industry, is less applicable to agriculture.

Sixthly, unlike in the manufacturing industry, combinations are not possible in farming due to the existence of a large number of small farm holdings. This results in acute competition among farmers.

Finally, in agriculture, farmers' control over production is very limited. When production is either not controlled or not controllable, serious maladjustments are likely to occur because at times more products will be produced than are required and could be sold profitably and at other times, not enough products are available when prices are high.

2

Economic Development and Agriculture

The importance of agriculture in the economic development of any country, rich or poor, is borne out by the fact that it is the primary sector of the economy which provides the basic ingredients necessary for the existence of mankind and also provides most of the raw materials which when transformed into finished products serve as basic necessities of the human race. In a preponderantly agrarian economy, agriculture plays a most strategic role from several points of view. At a minimum, farm production must be increased rapidly enough to keep pace with population growth. However, in a speedy industrialising economy, this is not enough. Industrialisation necessarily brings with it urbanisation and a rapid expansion of the industrial labour force. This may then be expected to bring with it a rising per capita demand for food, based on higher urban incomes.

In addition to supplying food, agriculture must provide many of the raw materials for industry. For instance, the fate of textile industry will be crucially affected by the supply of raw cotton; leather goods industry will depend on the availability of hides and skins; food processing brew and tobacco manufactures will all be dependent on agricultural supplies. Therefore, the pace of advance in a wide range of consumer goods manufactures will be eventually affected by the pace of agricultural development.

In addition, agriculture must generate export surpluses in order to earn the foreign exchange with which to finance the import of capital goods and certain kinds of industrial raw materials.

However, agriculture is not only a supplier of goods for domestic and export needs but is also a supplier of production factors such as capital and labour. A rapidly expanding industrial sector necessarily draws some of its labour force from the rural areas. Moreover, in one form or another, agriculture is called upon to save and finance a significant part of the investment for an expansion of industrial plant, transport and other sectors as well.

Moreover, farmers constitute a bulwark for social stability. They are assumed to be more industrious, less greedy and unenamoured by urban glamour. As they are less ambitious, they lead a contented life. With their typical sense of contentment, cheerful natural environment, self-sufficiency, and close-knit happy family life, they are relatively immune and invulnerable to the appeals of radicalism revolution, foreign ideologies and violence. It is well known that agriculture represents not only a predominant occupation of people in underdeveloped countries but also a culture by itself firmly rooted in the institutional set up.

The ideas which formed the intellectual and emotional basis of agricultural fundamentalism have deep roots in human history. For, mankind throughout-most of its existence has been either pastoral or farm people earning livelihood directly from earth. The shift to industrialism and urban living is of recent origin almost a current development when considered in the correct historical perspective.

The Physiocrats

Early theoretical literature on the role of agriculture in economic development can be traced to as far back as the eighteenth century in the writings of the physiocrats. Their doctrine constitutes the beginning of agricultural fundamentalism. The physiocrats discarded the mercantilist belief that wealth and its increase were due to exchange. They transferred it to the sphere of production, the power of creating wealth and the surplus which might be available for accumulation. The central point of their analysis was the search for this surplus and they formulated three basic tenets.

The first principle of physiocracy is that agriculture is the only productive industry and consequently the source of all wealth for the economy. It was their view that only agriculture, turned out a "net product" over and above its cost of production. They believe that the amount of food consumed by the workers plus what is used as seed is on the average less than the amount of produce raised from the ground.

Hence, the labour in agriculture is only productive, all the other labour employed in industry and commerce is sterile because they produce no net product, but only change the form or the title to the wealth produced by agriculture. According to Turgot, "what the agriculturist's industry causes the earth to produce beyond his personal wants is the only fund for the salaries which all the other members of the society receive." The degree of higher productivity of labour which made surplus possible made its appearance in agriculture and this surplus is a gift attributable to the productivity of nature not to the productivity of labour.

The second tenet of physiocracy was the beneficial role of natural order. Agriculture, according to them, is in accordance with natural order deriving the fruits of earth as given by God, whereas the products of the art made by man who is powerless to create. The essential aspects of natural order were the right to enjoy the benefits of property, to exercise one's labour and to follow one's self-interest.

The third principle of physiocracy is *laissez-faire* policy that state should not interfere with economic activity. As agriculture only created surplus value in conformity with the benevolence of nature, any interference with the natural order will lessen its contribution. Economic liberty is necessary for agriculture to encourage the production of surplus which makes a prosperous economy possible. Thus, free trade within a nation and between nations is required.

Adam Smith

Importance of agriculture in economic development was recognised by the classical writer's too. In view of the essential similarity in the political and economic climate in which Adam Smith and the physiocrats lived, it is not surprising to find that the general outlook of Adam Smith was not essentially different from the views of the founders of the French political economy. Smith's application of naturalism, his optimism in the role of invisible hand and his attack on mercantilism made him a great champion of *laissez-faire* than the physiocrats.

Adam Smith stated that it was a mistake to retard commerce and industry for the sake of agriculture. Nevertheless, Adam Smith considered agriculture as more productive than commerce and industry, because the forces of nature labour along with man.

Adam Smith indulged in the contrast of natural generosity of farmers with the ugly self-interest of those engaged in commerce and industry, in spite of his belief in the principle of self-interest as fundamental force of society and as the chief motive in life. He also agreed that investment in agriculture was quite in accord with the general interest of society. Hence, he may be considered to be an agricultural fundamentalist.

It is now believed that Adam Smith basic growth model refers only to the agricultural sector. Food, according to Smith, is the conditional factor in the growth of an economy. In his system, technical improvement in agriculture is the pivotal point for sparking of development in other sectors of the economy.

Let us suppose that we have a given stock of corn output from the previous year and wages paid in terms of corn are fixed. If average labour productivity is given, we can determine the produce of the present period S. Let:

S_{t-1} = corn output of the previous year (or wage fund). w = average wage per worker per period. p = average labour productivity.

Then $$S_t = \frac{S_{t-1}}{w}.p$$

We have assumed here that whole produce of the previous year has been converted into wage fund. But in actual practice, a proportion of this produce may be consumed by unproductive labourers. Let (1 - k) be unproductive consumption of corn from the previous year. Then this year's corn output will be—

$$S_t = \frac{(S_{t-1})K}{w}.p$$

Growth rate of the economy (*g*), therefore, may be defined as:

$$g = \frac{(S_t - S_{t-1})}{S_{t-1}} = \frac{p/w(S_{t-1}).k - S_{t-1}}{S_{t-1}}$$

$$= (p/w)k - 1$$

Hence, in Adam Smith's growth model, growth rate depends upon the value of *p*, *w* and *k*. Out of these parameters, the value of *w* and *k* are determined institutionally. Value of *p* depends on the stock of capital invested and the level of technology. Hence, in Smith's system, technical improvement in agriculture is the pivotal point for sparking off development in other sectors of the economy.

According to Smith, when by the improvement and cultivation of land, the labour of one family can provide food for two families, the labour of half the society becomes sufficient to provide food for the whole. The other half, therefore, can be employed in providing other things or in satisfying the other wants and fancies of mankind. The creation of an agricultural surplus is *sine qua non* for generating demand for other goods and services which could be purchased with the excess supply of agricultural products. Therefore, every increase in the surplus brings about more specialisation in industry through the division of labour.

Ricardo

Ricardo had scientifically explained the idea of Adam Smith. Ricardo considers agriculture as the most important sector of the economy. The difficulty of providing food for an expanding population serves as the focal point for his entire analysis. The central problem for classical economists was the analysis of the overall movement of the economy through times involving changes in population, capital accumulation and technical progress.

According to Adam Smith, the objective of economic analysis was the understanding of the nature and the causes of the wealth of nations. But to David Ricardo, the principal problem of political economy was to determine the "laws which regulate the distribution of commodities among the classes of society". It was through the laws which regulate distributive shares that he was trying to build up a simple macroeconomic model of economic development.

In Ricardo's vision of economic society, there are three major groups of actors on the economic scene: landlcrds who provide land, capitalists who provide capital and workers who provide labour.

In the progress of society by means of an expansion in population and an accumulation of capital, there arises, according to Ricardo, an increasing scarcity of the most fertile types of land. In order to meet the rising demand for food, the successive employment of equal units of labour and capital on poorer grades of land (together with the more intensive use of labour and capital on better grades of land) brings diminishing returns in terms of agricultural output. As poorer lands are brought under cultivation and diminishing returns occur, competition among the capitalists for the better grades of land causes a portion of produce of the land to be transferred to the landlords in terms of rent.

The rate of wages, according to Ricardo, is determined by the cost of subsistence. His assumption is that labourers have to be paid a minimum subsistence wage (food and other necessaries) if the supply of labour in the long-run is to be kept intact. This implies that, as the population grows, wage rates in money terms must rise (because foodgrains prices have risen due to the extension of margin of cultivation). This, in turn, squeezes the profit rate in agriculture and manufacturing. A lower profit rate curtails the rate of capital accumulation and in turn, the rate of growth in national income declines.

Thus, the law of diminishing returns from land dominates the economic scene and governs the fortunes of all classes. Ricardo thus showed how shortage of land would set a limit upon the expansion of agriculture and, therefore, by implication, upon that of the economy generally.

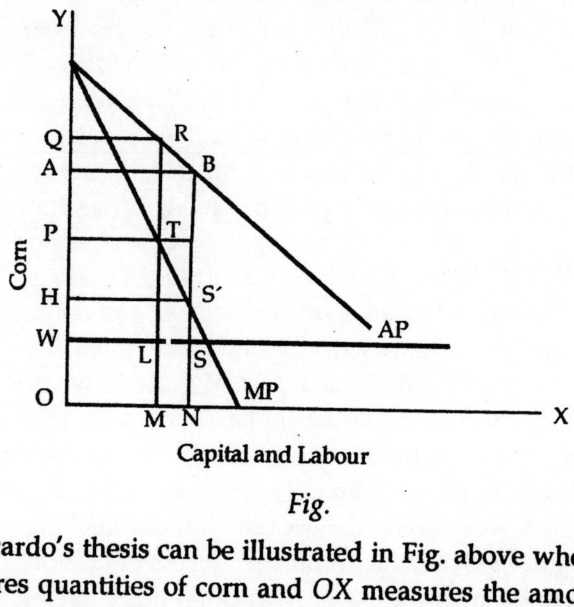

Fig.

Ricardo's thesis can be illustrated in Fig. above where *OY* measures quantities of corn and *OX* measures the amount of labour employed in agriculture. The curve *AP* represents the average product of labour and *MP*, the marginal product of labour. With *OM* amount of labour employed in agriculture sector, *OQRM* total corn is produces. Rent, according to Ricardo, is difference between *AP* and *MP*.

At the subsistence wage rate *OW*, the supply curve of labour *WL* is infinitely elastic, and the total wage bill is OWLM. Total profits WPTL are the residue after deducting rent and wages from the total produce, i.e.

$$WPTL = OQRM - (PQRT + OWLM)$$

Total output increases with economic development and with it the wage fund also rises. This leads to a proportionate increase in the amount of labour (population) which in turn increases the demand for corn which will result in rise of its price. As agricultural operations are subject to law of diminishing return rents continue to rise and profits, therefore, have a tendency to fall.

When amount of labour applied to agriculture sector increases from *OM* to *ON*, the total output increases to *OABN*. Out of this, *OWSN* is the total wage bill and *HABS* is the share of rent. The share of profit falls to *WLS'N*. This implies that as economic development proceeds, real wage-rate remains at the subsistence level and profits tend to fall. Obviously the change in national income goes in favour of landlords.

Limit on Growth

The Ricardian model of growth shows the accumulation capital, specially of corn surplus, as the main source of economic expansion. By analysing this aspect of growth, what Ricardo tried to show was how shortage of land would set a limit upon the expansion of agriculture and, therefore, by implication, upon that of economy generally. His two sector analysis of growth thus, shows how the operation of diminishing returns in agriculture will set a limit on the process of growth in non-agriculture by limiting the growth of working capital (wage goods) needed for the expansion of the later.

Let us suppose that total output (Y) is sum of agricultural and industrial outputs, i.e., Ya and Yi respectively. That is $Y = Ya + Yi$.

The Ricardian theory of distribution states that:

$$Ya = Wa + S \text{ and}$$

$$Yi = Wi + P$$

Where

Wa = Wage bill in agriculture,

S = Surplus Corn output.

Wi = Wage bill in industry.

P = total profit in industry.

Let us suppose that $Wi - S$ because the wage bill of industrial sector depends on surplus corn output in agriculture sector. Hence,

$$Yi = Ya - Wa + p \; (\because Ya - Wa = S)$$

Dividing by Y on both sides, we get:

$$\frac{Yi}{Y} = \frac{Ya - Wa + P}{Y}$$

$$\frac{Yi}{Y} \text{ (Share of industrial output)} = \frac{Ya}{Y} - \frac{Wa}{Y} + P/Y$$

The above equation shows that the share of industrial output in total output depends on the ratios of $\frac{Ya}{Y} - \frac{Wa}{Y}$ and $\frac{P}{Y}$. This also indicates that greater the ratio of $S/Y = \frac{(Ya - Wa)}{Y}$; greater the ratio of Yi/Y and greater the ratio of Ya/Y, greater the value of Yi/Y, assuming W/Y and P/Y as constant. An increase in Wa/Y will adversely affect the expansion of industrial output.

Thus, according to Ricardo, if agricultural output does not grow at a rate which is required to generate an amount of wage goods in order to cope with the demand of these goods for industrial workers, there is always tendency to raise money wage rate and as a consequence profit is bound to fall. According to Ricardo, "It is by the rise of price of corn that all other profits are regulated by agricultural profit. If the price of corn remained low, money wages would not rise and general profits could not fall."

It was only after Keynesian revolution in the 1930's that intellectuals started showing more interest in the theories of economic development. After the Second World War with many colonies winning their independence, literature on economic development started appearing. In the modern literature, role of agriculture in economic development has been further stressed.

Most of the underdeveloped countries who won their independence after the Second World War mostly suffered from three major features which impede their process of growth. The first feature is the existence of abundant supply of labour with marginal productivity of labour being negligible or approaching zero. The second is the underdeveloped nature of agriculture leading to low productivity of land. The third is the smallness of the capitalist sector with inadequate inevitable surplus for economic development.

Lewis Theory of Unlimited Supplies at Labour

Lewis in his well known article, "Economic Development Unlimited Supplies of Labour" presented a two sector model and investigated the expansion of the capitalist sector as it is fed by supplies of cheap labour from the agriculture sector. According to Lewis, under conditions, prevailing in majority of underdeveloped countries, the classical assumption of unlimited supply of labour is more relevant. Due to rapid rise in population in the countries, surplus labour exists in large sectors of economy where marginal productivity of such labour is negligible, zero or negative.

Lewis analyses the process of economic expansion in a dual economy composed of a "Capitalist" sector and a

"Subsistance" sector. Output per head in subsistance sector is lower than that in the capitalist sector. People in subsistance sector are generally backward, illiterate and unskilled with the result that their average productivity is low. On the other hand, people engaged in capitalist sector are advanced, literate and skilled.

The basic assumption of Lewis', Model is that there exists surplus labour in the subsistance sector. Such labour is there either with zero marginal productivity or having marginal productivity much less than the institutional wage. According to Lewis, the surplus labour in the subsistance sector acts as a source from which an unlimited supply of labour can be drawn for the development of the capitalist sector. "In this situation, new industries can be created or old industries can be expanded without limit at the existing wage, or to put it more exactly, shortage of labour is no limit to the creation of new sources of employment".

In Lewis' model of growth, savings play a crucial role. In his model, if capitalists do not reinvest a larger and larger proportion of their profits, neither will the total product expand nor will opportunities for employment increase. "The central problem in the theory of economic development is to understand the process by which a community has previously been saving and investing 4 to 5 per cent of its National income or less, converts itself into an economy, where voluntary saving is running about 12 to 15 per cent of national income or more.

As explained above, surplus labour exists in the subsistance sector and it can be attracted to the industrial sector if it is offered a wage rate which is a little higher than the institutional wage rate prevailing in the rural areas. Capitalist wage is generally higher than the subsistance earnings in order to compensate labour for the cost of transferring and to induce labour to leave the traditional life of the subsistance sector. At the existing capitalist wage, however, the supply of labour is considered to be perfectly elastic. This situation has been illustrated in the following Fig.

In this figure, *OW* is the real wage rate in the capitalist sector and *WS*, the perfectly elastic supply of labour. Given a fixed amount of capital at the outset, the demand for labour is initially represented by the marginal productivity schedule of labour, N_1 D_1. If OW is the current wage rate, the amount of labour employed in the capitalist sector is *OL*, and beyond *L*, workers earn whatever they can in the subsistence sector. The total product N_1 *PLO* in the capitalist sector will then be divided between wages in the amount *OWPL* and the capitalist's surplus in the amount WPN_1.

Quantity of Labour

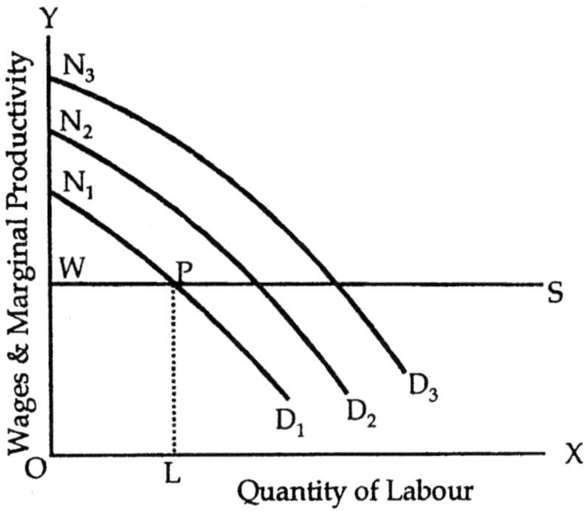

Fig.

In tracing the process of economic expansion, Lewis emphasises that the key to the process is the reinvestment of the capitalist system. As the capitalist sector expands, labour is withdrawn from the subsistence sector into wage employment, the surplus then becomes even larger, there is still more reinvestment of profits, and the process continues on, progressively absorbing the surplus labour from the subsistence sector.

The process of capital formation, as envisaged by Lewis, comes to an end when capital accumulation catches up with population, so that there is no longer surplus labour in the subsistence sector left to be absorbed in the industrial sector. Beyond this point, real wages no longer remain constant, but instead, rise as capital formation occurs. It will now be in the interest of the producers in the subsistence sector to compete for labour since the marginal product of labour will no longer be below the institutional wage. When this point is reached, the agricultural sector is said to have been commercialised.

Fei and Ranis Model

Attention needs to be drawn to the fact that both Nurkse and Lewis recognised the importance of agriculture in providing "wage goods" to the industrial workers. Failure to provide food could seriously limit the process of economic expansion envisaged by their models. Yet they did not seriously prescribe a clear and suitable line of action which could keep off such a danger to the process of economic development. Fei and Ranis are too careful to miss laying stress on the strategic role that agriculture is desired to play in the entire development effort.

In their view: "The strategic nature of agricultural sector in the dualistic economy is due not only to its preponderant size but also .... to the peculiar production and consumption conditions attached to the agricultural goods."

Unlike Nurkse and Lewis, agricultural innovations and austerity are brought to the forefront of the development effort, given due recognition and put in their proper perspective by Fei and Ranis.

The Fei-Ranis model also focuses attention on the transfer of labour from the agricultural to the non-agricultural sector as central to economic development. Major and well-recognised social problem of underdeveloped countries, according to Fei and Ranis, is the existence of the so-called overpopulation in its agriculture sector. This denotes the persistent pressure of population against scarce resources, mainly land and thereby leading to the worsening land-labour ratio. Classical growth model and the pessimistic conclusions of Ricardo and Malthus based on the assumption of stationary technology and the law of diminishing returns in the agricultural sector has been depicted by Fei and Ranis in Fig. below.

In bellow Fig.; the inputs used are labour and capital which have been measured on horizontal and vertical axes respectively. Production contour lines are represented by curves M, M′, M″ .... For the sake of simplification, constant returns have been assumed. Ridge lines OV^* and OU^* mark off the region of factor substitutability, *e.g.*, below OV^*, the production contour lines become horizontal indicating that with land held constant, any further increase in labour renders that factor redundant as output can no longer be increased. If the total amount of land is supposed to be Ot, the amount of labour which can be absorbed without becoming redundant can be determined by the ridge line, ts.

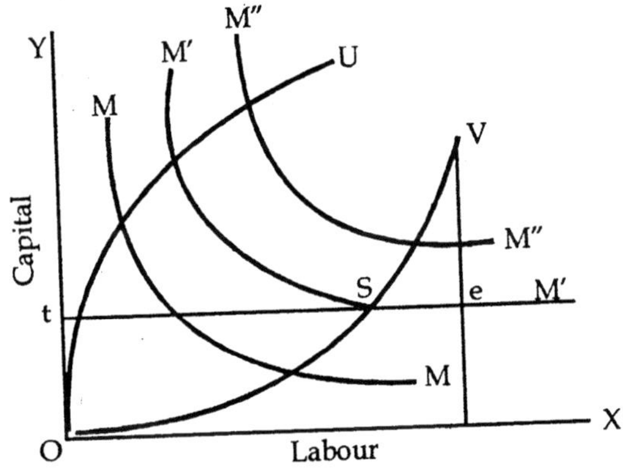

Fig.

Therefore, the labour utilisation ratio $R = ts/ot$ (labour which can be employed per unit of land productively) which is the slope (inverted) of the ridge line OV^*. The population density (number of workers per unit land) in the agricultural sector, called the endowment ratio S = *te/ot*. The labour force which is productive is *ts* and the surplus labour is *se*. The fraction of the existing agricultural labour force which is productive is called the non-redundancy co-efficient *T*, which is equal to:

$$\frac{ts}{te} = \frac{ts/to}{te/ct} = \frac{R}{S}$$

Non-redundancy radio *T*, therefore, is directly proportional to the labour utilisation ratio *R* and inversely proportional to the endowment ratio *S*. It describes how favourably a given economy is endowed with arable land relative to its agricultural population under the existing conditions of production techniques.

When the amount of arable land is given, the production conditions described above can be pictured alternatively by familiar curves representing total physical productivity of labour (TPP_L) and marginal physical productivity of labour (MPP_L).

Fig. shows TPP_L and MPP_L curves. It is clear from Fig. that TPP_L increases at a decreasing rate when more and more labourers are added to the fixed amount of land *ot*, until it becomes horizontal at M. Similarly, the curve MPP_L gradually decreases as the law of diminishing returns is making its effect felt, until at point G, the MPP_L becomes zero.

In the context of Fig., the non-redundancy co-efficient *T* is equal to *OQ/OP*.

From Fig., average productivity *APP* = *MP/OP*

The institutionally determined real wage level is set near subsistence and is related to the average productivity of agricultural labour APP_L. This is called the constant institutional wage (CIW).

From Fig. bellow, the portion of redundant labour force in the economy = *PQ*. If out of this redundant labour force, *PY* is allocated to the industrial sector, the remaining labour force in agricultural sector produces an output of *YZ* units, while at constant institutional wage, its total real income is represented by *XY* units. The difference *XZ* represents the agricultural surplus. This total agricultural surplus *TAS* emerges as a result of the allocation of the redundant labour force *PY* out of the agricultural sector. Redundant labour force in agricultural sector disappears at *Q* and at this point, the disguised unemployed agricultural labour force disappears and commercialisation of agriculture occurs.

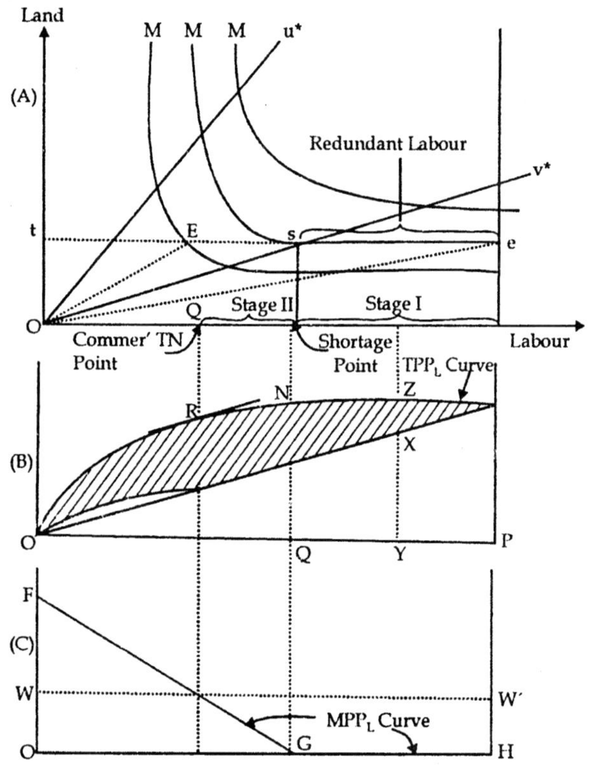

Fig.

In order to have a total view of development process in the context of dualistic economy, we now turn to the industrial sector. The role of industrial sector is essentially related to the expansion of employment opportunities required for the absorption of the surplus labour force released by the agricultural sector and thereby a gradual expansion of industrial productive capacity and output.

Diagrammatic Exposition

The process of labour absorption by the industrial sector of the dualistic economy has been exhibited diagrammatically in the previous figure. In the double deck diagram given below, upper deck represents the production contour map of the industrial sector.

The production contour lines indexed by Q_0, Q_1, Q_2 are once again assumed to exhibit the property of constant returns to scale. The central expansionary role of the industrial sector may be symbolically represented by an expansion path through time A_0, A_1, A_2, representing a gradual expansion of the capital stock K_0, K_1, K_2, of the industrial labour force L_0, L_1, L_2, and industrial output Q_0, Q_1, Q_2.

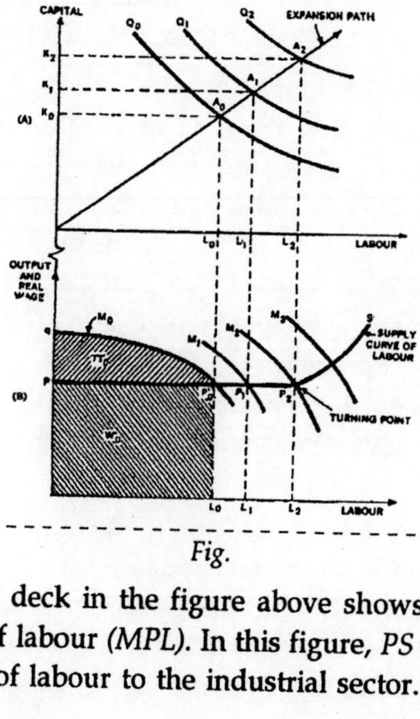

Fig.

The lower deck in the figure above shows the marginal productivity of labour *(MPL)*. In this figure, *PS* represents the supply curve of labour to the industrial sector. Its horizontal portion PP_2 corresponds to the existence of the pool of redundant labour which can be absorbed by the industrial sector at a constant wage W, which is greater than W, the subsistence institutional wage. The rising portion *PS* curve is marked by a turning point P_2, which corresponds to the situation when redundant labour is completely absorbed by the industrial sector.

With respect to the industrial sector of the dualistic economy is reasonable to accept the customary competitive assumptions in the labour market implying that industrial entrepreneurial behaviour is characterised by profit maximisation. Under this assumption, the marginal productivity of labour takes an operational significance as the industrial demand curve for labour.

For each amount of capital stock K_0, K_1, K_2, the marginal physical productivity of labour curve *(PPL)* may then be drawn, that is, curves M_0, M_1, M_2. The competitive employment equilibrium positions in the labour market are consequently indicated by the intersections of these curves with the industrial supply curve, *i.e.*, points P_0, P_1, P_2.

Let us suppose that there exists a stock of real capital goods of volume K_0 in the industrial sector. The equilibrium employment position is then determined by curve M_0 and indicated by the point P_0. At this point, the industrial sector is making profit (represented by the shaded area *PPO)* and the magnitude of total real wage income is W_0 $(OL_0\ P_0P)$. At the low level of per capita income, it is reasonable to assume that workers do not save but industrial profits (II_0) constitute the major source of investment funds originating in the industrial sector.

Moreover, in determining the total supply of investment funds available for industrial capital accumulation, it should be noted that there exists a second important source, the hidden rural savings. If these savings (say) so are siphoned off to the industrial sector, then the total investment fund of the industrial sector is $II_0 + S_0$, SO that the total capital stock in the next period will be K_1

With this new capital stock K_1, a new MPP_t curve is determined, $M_1 0$. This determines a new equilibrium position P_1 causing an increase in employment of labour by the amount $L_0 L_1$. This represents an additional transfer of labour from the agricultural sector to the industrial sector as a result of capital accumulation.

In this fashion, we have shown how the agricultural sector constitutes an important source of supply of both manpower and savings for fuelling the expansion of the industrial sector. According to the above described process, industrialisation and output growth may be viewed as a continuous shifting of the MPP_L curve to the right ($M_0 M_1 M_2$) through time. Associated with such a shift, there must occur a continuous process of reinvestment of industrial profits (II_0 II_1 II_2) and of the channelisation of agricultural surplus (S_0 S_1 S_2) into the industrial sector in order to finance a continuous expansion of industrial capital stock (K_0 K_1 K_2). This, in turn, leads to a continuous increase in the demand for, and employment of, labour (L_0 L_1 L_2) and a continuous expansion of industrial output (Q_0 Q_1 Q_2)

In this manner, we have briefly outlined the crucial significance of the agricultural sector in the economic development, in that it constitutes an important source of supply of both manpower and savings for fuelling the expansion of the industrial sector. Disguised unemployment in over populated economies responds to stimulants of economic growth by providing cheap labour at constant real wages for non-farm sectors. During this phase, it is plausible to improve agricultural productivity by relaxing the constant real wage assumption. The development of agriculture raises the rural purchasing power and provides a mass market for industrial goods and results in the expansion of investment opportunities.

It is, thus, obvious from the contribution of Lewis and Ranis and Fei that mobility of labour from the farm to the non-farm sector is an essential ingredient of economic transformation as well as development of agriculture.

Experience of all the developed economies indicates that due to the operation of various pull and push factors contributing to the migration of labour from rural to urban areas, the share of agricultural sector in the total labour force of the country diminishes. Consequent upon the transfer of labour from the agricultural sector, the productivity of labour and capital in the farm sector increases and this leads to an increase in the earnings of farmers. Increased earnings of farmers create essential conditions for the adoption and absorption of new agricultural strategy, which in turn contributes to enhanced returns per acre.

It is now customary to summarise in four ways how greater agricultural productivity and production contribute to an economy's development:

1. It helps in faster development by supplying foodstuffs to the rapidly increasing population and raw materials to other expanding sectors in the economy;
2. The developing agricultural sector also provides an investible surplus of savings and taxes to support investment in another expanding sector;
3. The rising agricultural income of the agricultural sector will raise the demand of the rural population for products of other expanding sectors;
4. It can also contribute to the economic development of an economy by relaxing the foreign-exchange constraint by earning foreign exchange through export or by saving foreign exchange through import substitution.

Kuznet summarises these contributions as:

1. The product contribution.
2. The factor contribution.
3. The market contribution.

Precisely, the most important contribution of agriculture to the economic development is that constituted by growth of product within the sector itself. An increase in the net output of agriculture, in and of itself, represents a rise in the product of the country – since the latter is the sum of the increases in the net products of the several sectors.

This product contribution can be examined firstly as a contribution to the growth of total net or gross product, and secondly to the growth of product per capita. Algebraic notation of the above argument is given below:

Pa = Product of agriculture sector (Sector A)

Pb = Product of all other sectors (Non-Agriculture Sector B)

P = Total product = $Pa + Pb$

Δp = Increment in total product

Y_a = rate of growth of Pa

Y_b = rate of growth of Pb, so that

$Pa^1 = Pa^0 (1+Y_a)$, the subscripts referring to time

$Pb^1 = pb^0 \ (1+Y_b)$

Then, $AP = Par_a + Pbr_b$

The equation for the share of the growth of agriculture product in the growth of total product is therefore:

$$t = \frac{P_a r_a}{\Delta P} = \frac{1}{1+\left(\frac{P_b}{P_a} \times \frac{\gamma_b}{r_a}\right)}$$

Thus, according to Kuznetz; if at the initial point of time, the share of agricultural sector in countrywide product is 60 per cent – and if over the next decade the rate of growth of the no – A sector *(rb)* is four times as high as that of the A sector *(ra)*, the product contribution of agriculture to the growth of total product will be one divided by [1+ (0.67 × .4)] or about a quarter.

The product contribution of agriculture towards an overall economic development takes two forms.

These are:

Provision of Wage Goods: Economic development is characterised by a substantial increase in the demand for agricultural products and failure to expand food supplies in pace with the growth of demand can seriously impede economic growth. Developing phase of an underdeveloped economy results in *(i)* increased population, *(ii)* Shifting of labour from rural areas to urban areas, *(iii)* increase in per capita income. All these changes would lead to higher demand for foodstuffs.

The annual rate of increase in demand for food is given by $D = P + ng$, where P and g are the rate of growth of population and per capita income and $r\backslash$ is the income elasticity of demand for agricultural products.

As indicated by Johnston and Mellor, not only are there high rates of population growth in *LDCs* but the income elasticity of demand for food in these countries is considerably higher.

If food supplies fail to expand in pace with the growth of demand, the result is likely to be a substantial rise in food prices leading to political discontent and pressure on wage rates with consequent adverse effects on industrial profits, investment, and economic growth.

Provision of Industrial Raw Material

Economic history of most of the advanced countries will show that the agro-based industries were first to develop in such countries. These industries which draw their basic raw material from agricultural sector will flourish only if a continuous supply of such raw material is made available to them. In the first phase of industrialisation of an economy, agro-based industries get priority firstly because it is easier to produce raw materials in the agricultural sector. And secondly such industries can be started with traditional technology. It is also easy to shift labour from agricultural sector to such factories. To conclude, the industrial development in the initial stages requires that more of its raw materials have to be produced in the agricultural sector.

Market Contribution

According to Kuznetz, market contribution of a sector takes place when "the given sector provides such opportunities by offering part of its product on domestic or foreign markets in exchange for goods produced by the other sectors, at home or abroad".

Thus, agriculture makes a market contribution to economic growth by (1) "Purchasing some production items from other sectors at home or abroad; (2) Selling some of its products not only to pay for the purchases listed under (1) but also to purchase consumer goods from other sectors or from abroad or to dispose of the product in any way other than consumption within the sector. In all these ways, agriculture makes it feasible for other sectors to emerge and grow and for international flows to develop; just as these other sectors and the international flows make it feasible for the agricultural sector to operate more efficiently as a producing unit and use its product more effectively as a consuming unit".

A developed agriculture not only enter the market to dispose off its marketed surplus but also makes market contribution by making heavy purchases of modern inputs required for its own development. The forward and backward linkages of developed agriculture thus help in the establishment of most of the industries using either agricultural raw material or supplying agricultural inputs. According to David Metcaff, agricultural development through providing market for agricultural inputs and consumer goods, promotes the development of industrial sector. Developed industrial sector, in turn, helps in the development of the agricultural sector through the speed of modern technology in agriculture and by providing an expanded market for agricultural products. This is a virtuous circle which in the process gives rise to institutions facilitating two way exchange of commodities.

Developed agriculture can also contribute to the development of international trade. According to Johnston and Mellor, "expansion of agricultural exports is likely to be one of the most promising means of increasing incomes and augmenting foreign exchange earnings in a country stepping up its development efforts". According to Kuznetz, "Since agriculture, after mining, is the sector in which natural endowments have greatest weight, it is hardly a surprise that in the initial stages of growth of many presently developed countries, agriculture was a major source of exports and that the resulting command over the resources of the more developed countries played a strategic role in facilitating modern economic growth".

Factor Contribution

Factor contribution of agriculture ot economic growth occurs when some resources are transferred from agriculture to non-agricultural sectors of the economy. Two most important factors which a developing agricultural sector contributes towards the development of other non-agricultural sectors are capital and labour.

Capital Contribution

An underdeveloped country that is making determined efforts to achieve economic progress faces formidable requirements for capital to finance the creation and expansion of manufacturing enterprises. These requirements are certain to outstrip the supply of funds in most of the underdeveloped countries. "Since there is scope for raising productivity in agriculture by means that require only moderate capital outlays, it is possible for the agricultural sector to make a net contribution to the capital requirements for infrastructure and for industrial expansion without reducing the low levels of consumption characteristic of the farm population in an underdeveloped country".

According to Mellor, the contribution of the agricultural sector to capital formation may be marshalled in four ways:

1. It may be extracted by the government through the medium of taxes such as land tax, agricultural income tax, etc.
2. Agricultural production may be increased sufficiently to bring about a relative decline in agricultural prices and hence favour increased profits in the non-farm sector which in turn bring about favourable effects on savings and investment in that sector;
3. Agriculture may form capital directly within its own sector and minimise its own demand for capital from other sectors;
4. Agriculture may invest directly in other sectors – perhaps after its own development has increased demand for products from other sectors.

The transfer of capital from agricultural sector to other non-agricultural sectors may be voluntary or compulsory. Compulsory transfer of funds from agricultural sector for the benefit of other sectors is ordinarily done through taxation in which the burden on agriculture is far greater than the services rendered by the government to agriculture. This kind of tax has played a significant role in the early development of Japan, England and Russia.

In Japan, for example, in the last two decades of the 19th century, the land tax was over 80 per cent of the central government taxation. Forced extraction of surplus from agriculture by taxation or arbitrarily keeping low prices of agricultural products have been other measures taken by different governments to transfer funds from agricultural sector to non-agricultural sector.

On the voluntary basis, the farm sector may, of its own, lend or invest its savings in the growth of the non-agricultural sectors. But in the absence of adequate data, it is difficult to measure the extent to which savings originating in agriculture sector contribute to the financing of capital formation elsewhere in the economy.

In the absence of such data we can only speculate on such magnitudes. However, the share of domestic savings originating in agriculture is function of *(i)* the share of agriculture in total income, *(ii)* the lower level of real income in agriculture than in the total sectors, and, *(iii)* the relative propensity to save of the agricultural population and of other groups in the economy.

Nurkse has referred to another way through which agriculture can provide capital to non-agricultural sectors. Another way of transferring resources from agricultural to non-agricultural sectors is by the government turning the terms of trade against agriculture by imposing price controls on agricultural products.

If the improvement in the terms of trade in the non-agricultural sectors raises their incomes, and beneficiaries save at a higher marginal rate than the decreased agricultural incomes, aggregate saving rates will increase and thus, agriculture will have made a net contribution to total savings in an indirect manner.

Labour Contribution

Another important factor contribution of agricultural sector to the non-agricultural sector is the provision of labour. In most underdeveloped economies, due to rapid rise in population and absence of manufacturing sector, agricultural sector sustains more labour than required. This type of labour is known as disguised labour. With the overall development of the economy, labour migration takes place from agricultural sector to non-agricultural sector.

In the earlier stages of the development of an underdeveloped economy, the bulk of the labour for the expanding sectors must be drawn from agriculture simply because there is almost no other source.

Kuznetzs has expressed the importance of transfer of labour from the agricultural sector to non-agricultural sectors in another way. According to him, this transfer of workers from agricultural sector to non-agricultural sector means a sizeable capital contribution because each migrant is of working age and represents some investment in past rearing and training to maturity.

This factor contribution of the agricultural sector to non-agricultural sectors must have been quite large in the early and even later phases of modern economic growth.

Agricultural Fundamentalism

A group of thinkers (Barker, Wilson Barsodi and Humphries, etc.) have commonly held the orthodox view that agriculture is par excellence a fundamental industry.

They support this contention with the explanation that agriculture being the producer of basic food for the human beings is the basis of existence of the human race. Besides, it produces raw materials for many industries and also is an important constituent of trade. Hence, it is a foundation of manufacture as well as commerce.

Adam Smith in his *Wealth of Nations* took vigorous exception to the physiocrats' notion that agriculture is the only productive activity, but still he ranked agriculture as the highest in wealth-creating powers. 1798, laid a very pronounce emphasis on agriculture.

T.R. Mathus, in his first easily on population, which appeared in 1978, laid a very pronounce emphasis on agriculture. The fate of capitalistic economy, according to Ricardo, purely depended upon the yield rate in the agricultural sector.

Henry Carey and his father, Mathew Carey, though strong exponents of industrialisation, still affirmed the overwhelming importance of a prosperous agriculture.

To these beliefs that agriculture is the basic industry, and any philosophy that may be invented to go with it, Dr. S. Davis of the food Research Institute, in an essay published in 1935 gave the name, "Agricultural Fundamentalism".

This view has been challenged by another group of modern thinkers. T.S. Davis, who is one of the most popular representatives of this group is of the view that "the wealth and welfare of nations depend upon many complex conditions. Today, agriculture is not uniquely basic and the prosperity of a nation depends largely on other factors than the work of those who till the soil". This view has been supported by the argument that with the progress of any society, the relative importance of agriculture had always declined.

Both the above views stand poles apart. A more balanced view has been taken by the economists like Karl Brandt, H.R. Tolly and P. Chew who assign the agricultural sector a status equal to that of other sectors of economy. K. Brandt holds that, "farmers are a vital part of the arterial system of circulation through which flow the goods and services of the national economy.

The nation depends on properly functioning farms as an important source of primary materials, food, and fibres, yet the farms cannot be treated as an independent object of policies, nor can they be made prosperous in emancipation from the remainder of the economy. Nor can the conditions creating mass unemployment and decreased output in cities be cured by maintaining or restoring economic well-being to the farmers alone."

A still better explanation has been held by the writers like Shultz, W.W. Wilcox and others. They believe that there is interdependence and close relationship of the various sectors of economy. In this connection, Wilcox says, "when we say that it is any sense more a generator of income in modern society than other occupations, we fail to understand the true nature of other occupations, we fail to understand the true nature of modern economic society.

An efficient agriculture made up of farm families with a high standard of living and a high buying power per person contributes much towards a high national income and the economic "well-being of the nation, but the same can be said for each of the other groups it is impossible to say which is most important in modern economic society".

Declining Importance of Agriculture

In the developing countries which are predominantly agricultural, the share of agriculture in national income is quite substantial. As the country starts developing, there is gradual decline in the contribution made by agriculture to national income.

As the economy grows and industrial sector develops, two things happen (1) importance of the agricultural sector as the one which initiates and sustains economic growth starts declining; and (2) importance of agriculture sector as the premier sector of production also starts declining.

At the higher stage of development, as industrial sector develops, its dependence on the agricultural sector for the provision of the various factors of production as well as raw materials declines. Industrial sector starts generating its own savings and thus capital begins to be supplied by the industrial sector itself.

The dependence of industrial sector for agricultural raw materials also declines since new technology makes it possible to develop mining industry and many mining based industries come into being. Similarly need for labour also is cut down since capital intensive technology replaces labour intensive technology.

With the development of the economy, Agriculture sector also loses its importance as the main source of national income. In the most highly developed countries, the contribution of agricultural sector to their national income has declined to very low level as is clear from the following table.

Table: Originating from Agriculture

Country	*Percentage of Working Population Engaged in Agriculture*	*Percentage of Gross Domestic Product (at factor cost)*
Canada	5.2	3.4
Denmark	8.3	5.7
France	8.4	4.6
Japan	9.7	3.4
UK	2.7	2.0
USA	3.6	1.6
India	55.4	31.4

Prof. Edward Nissan in his recent study has tried to show this fact by providing a simple empirical measurement for various countries. He has drawn data from various World Bank documents (1984-92).

Prof. Nissan has adopted the World Bank division of the countries into *(i)* low income, *(ii)* lower-middle income, *(iii)* Upper-middle income, and *(iv)* high income, according to levels of income.

Model

Edward Nissan has tried to measure the magnitude of change of the non-agricultural sector to the agricultural sector with the help of a procedure suggested by Ghatak and Ingersent (1984) (Agricultural and Economy Development)

Let P_n and P_a be the gross products of the agricultural and the non-agricultural sect respectively. Then P_n/P_a is a measure of the magnitude of the non-agricultural sector compared to the agricultural sector.

Let the symbol "0" and "1" stand for two distinct periods of time. The magnitude of change in the ratio of the non-agricultural sector to the agricultural sector over this period is approximated by:

$$\Delta(P_n / P_a) = (P_n^1 / P_a^1) - (P_n^0 / P_a^0) \quad ...(1)$$

To discover the meaning embodied in equation (1) assume that the period "1" corresponds to t years after an initial period "0".

Then:

$$P_a^1 = P_a^1 (1 + r_a)^t \text{ and } P_n^1 = P_n^0 (1 + v_n)^t$$

Where r_a and r n are the average annual growth rates between the two periods of time. By substitution and simplification in (1)

$$\Delta(P_n / P_a) = P_n^0 \left[(1 + r_n)^t - (1 + r_a)^t\right] / \left[(P_a^0 (1 + r_a)^t\right] \quad ...(2)$$

Equation (2) is of name help in deducing relationship between the agricultural sectors and non-agricultural sectors. For instance, the magnitude is determined by initial ratio (P_n^0 / P_a^0), and the compound growth of the two sectors $(1+ r_n)^t$ and $(1 + r_a)$ If $r_n > r_a$ and P_n^0 / P_a^0 is relatively large, then the magnitude of change is positive and large.

The major points of interest in this study are the ratio (P_n/P_0) and the magnitude of change $\Delta(P_n / P_a)$ over a period of time. For purpose of comparisons, three periods of time were chosen; 1965, 1984 and 1990. Results of computations of the ratio's P_n / P_a are presented in the following table which tells for each country how many folds larger the non-agricultural sector is than the agricultural sector. Some countries display small magnitudes in 1984 or 1990 or both than in an earlier period.

From the following table, the largest change was witnessed by Hong Kong and Singapore where the ratios in 1984 and 1990 of almost 100 are the largest than those in 1965. For United States and the United Kingdom, the ratio increased from 33 in 1965 to 49 in both 1984 and 1990.

The following table summarises information and shows for each year the mean of (P_n/P_a), expressed as (x), the standard deviation (s), the co-efficient of variation *(cv)* obtained as (S/X) and the minimum and maximum of observations of each of the four groups of economies. As evident from the following table, a systematic change in the average X is apparent both in the direction of time from 1965 to 1990 and in the direction of economies from low income to high income.

In both directions the average magnitude (P_n/P_0) increases – exactly what is expected from economic development theories which proclaim that the agricultural sector shrinks overtime as compared to the other sectors for every country. These theories also hold that as the income rise, the importance of the agricultural sector diminishes.

Prof. Edward summarises his study in the following lines. He writes, "the economy of each nation is made up of three broad sectors; agricultural, industrial and services, while in developing countries the share of the agricultural is large, through the process of development the agricultural sector shrinks in favour of the other sectors. Therefore, one may use the criteria of the ratio of the non-agricultural to the agricultural sector as a means to categorise the level of development of a country.

The World Bank has classified important economies of the world, in a rational manner. The following table shows it well.

Significant Economies (World Bank List — 2009)

Economy	*Code*	*Region*	*Income Group*
Afghanistan	AFG	South Asia	Low income
Algeria	DZA	Middle East & North Africa	Upper middle income
Angola	AGO	Sub-Saharan Africa	Lower middle income
Argentina	ARG	Latin America & Caribbean	Upper middle income
Aruba	ABW	..	High income: nonOECD
Australia	AUS	..	High income: OECD
Austria	AUT	..	High income: OECD
Azerbaijan	AZE	Europe & Central Asia	Lower middle income
Bahrain	BHR	..	High income: nonOECD
Bangladesh	BGD	South Asia	Low income
Belgium	BEL	..	High income: OECD
Bhutan	BTN	South Asia	Lower middle income
Botswana	BWA	Sub-Saharan Africa	Upper middle income
Brazil	BRA	Latin America & Caribbean	Upper middle income
Brunei Darussalam	BRN	..	High income: nonOECD
Cambodia	KHM	East Asia & Pacific	Low income
Canada	CAN	..	High income: OECD
Central African Republic	CAF	Sub-Saharan Africa	Low income
Chad	TCD	Sub-Saharan Africa	Low income
Chile	CHL	Latin America & Caribbean	Upper middle income
China	CHN	East Asia & Pacific	Lower middle income
Colombia	COL	Latin America & Caribbean	Upper middle income
Comoros	COM	Sub-Saharan Africa	Low income
Congo, Dem. Rep.	ZAR	Sub-Saharan Africa	Low income
Congo, Rep.	COG	Sub-Saharan Africa	Lower middle income
Cuba	CUB	Latin America & Caribbean	Upper middle income
Cyprus	CYP	..	High income: nonOECD
Czech Republic	CZE	..	High income: OECD
Denmark	DNK	..	High income: OECD
Ecuador	ECU	Latin America & Caribbean	Lower middle income
Egypt, Arab Rep.	EGY	Middle East & North Africa	Lower middle income
El Salvador	SLV	Latin America & Caribbean	Lower middle income
Eritrea	ERI	Sub-Saharan Africa	Low income
Estonia	EST	..	High income: nonOECD
Ethiopia	ETH	Sub-Saharan Africa	Low income
Fiji	FJI	East Asia & Pacific	Upper middle income
Finland	FIN	..	High income: OECD
France	FRA	..	High income: OECD
Gabon	GAB	Sub-Saharan Africa	Upper middle income
Gambia, The	GMB	Sub-Saharan Africa	Low income
Georgia	GEO	Europe & Central Asia	Lower middle income
Germany	DEU	..	High income: OECD
Ghana	GHA	Sub-Saharan Africa	Low income
Greece	GRC	..	High income: OECD
Greenland	GRL	..	High income: nonOECD
Guatemala	GTM	Latin America & Caribbean	Lower middle income
Guinea	GIN	Sub-Saharan Africa	Low income
Guinea-Bissau	GNB	Sub-Saharan Africa	Low income
Guyana	GUY	Latin America & Caribbean	Lower middle income

Contd...

Economy	*Code*	*Region*	*Income Group*
Haiti	HTI	Latin America & Caribbean	Low income
Honduras	HND	Latin America & Caribbean	Lower middle income
Hong Kong, China	HKG	..	High income: nonOECD
Hungary	HUN	..	High income: OECD
Iceland	ISL	..	High income: OECD
India	IND	South Asia	Lower middle income
Indonesia	IDN	East Asia & Pacific	Lower middle income
Iran, Islamic Rep.	IRN	Middle East & North Africa	Lower middle income
Iraq	IRQ	Middle East & North Africa	Lower middle income
Ireland	IRL	..	High income: OECD
Israel	ISR	..	High income: nonOECD
Italy	ITA	..	High income: OECD
Jamaica	JAM	Latin America & Caribbean	Upper middle income
Japan	JPN	..	High income: OECD
Jordan	JOR	Middle East & North Africa	Lower middle income
Kazakhstan	KAZ	Europe & Central Asia	Upper middle income
Kenya	KEN	Sub-Saharan Africa	Low income
Korea, Dem. Rep.	PRK	East Asia & Pacific	Low income
Korea, Rep.	KOR	..	High income: OECD
Kuwait	KWT	..	High income: nonOECD
Kyrgyz Republic	KGZ	Europe & Central Asia	Low income
Latvia	LVA	Europe & Central Asia	Upper middle income
Lebanon	LBN	Middle East & North Africa	Upper middle income
Liberia	LBR	Sub-Saharan Africa	Low income
Libya	LBY	Middle East & North Africa	Upper middle income
Lithuania	LTU	Europe & Central Asia	Upper middle income
Luxembourg	LUX	..	High income: OECD
Macao, China	MAC	..	High income: nonOECD
Macedonia, FYR	MKD	Europe & Central Asia	Upper middle income
Madagascar	MDG	Sub-Saharan Africa	Low income
Malawi	MWI	Sub-Saharan Africa	Low income
Malaysia	MYS	East Asia & Pacific	Upper middle income
Maldives	MDV	South Asia	Lower middle income
Mali	MLI	Sub-Saharan Africa	Low income
Malta	MLT	..	High income: nonOECD
Mauritania	MRT	Sub-Saharan Africa	Low income
Mauritius	MUS	Sub-Saharan Africa	Upper middle income
Mayotte	MYT	Sub-Saharan Africa	Upper middle income
Mexico	MEX	Latin America & Caribbean	Upper middle income
Moldova	MDA	Europe & Central Asia	Lower middle income
Monaco	MCO	..	High income: nonOECD
Mongolia	MNG	East Asia & Pacific	Lower middle income
Montenegro	MNE	Europe & Central Asia	Upper middle income
Morocco	MAR	Middle East & North Africa	Lower middle income
Mozambique	MOZ	Sub-Saharan Africa	Low income
Myanmar	MMR	East Asia & Pacific	Low income
Namibia	NAM	Sub-Saharan Africa	Upper middle income
Nepal	NPL	South Asia	Low income
Netherlands	NLD	..	High income: OECD
New Zealand	NZL	..	High income: OECD
Nicaragua	NIC	Latin America & Caribbean	Lower middle income

Contd...

Economy	*Code*	*Region*	*Income Group*
Niger	NER	Sub-Saharan Africa	Low income
Nigeria	NGA	Sub-Saharan Africa	Lower middle income
Norway	NOR	..	High income: OECD
Oman	OMN	..	High income: nonOECD
Pakistan	PAK	South Asia	Lower middle income
Panama	PAN	Latin America & Caribbean	Upper middle income
Paraguay	PRY	Latin America & Caribbean	Lower middle income
Peru	PER	Latin America & Caribbean	Upper middle income
Philippines	PHL	East Asia & Pacific	Lower middle income
Poland	POL	Europe & Central Asia	Upper middle income
Portugal	PRT	..	High income: OECD
Qatar	QAT	..	High income: nonOECD
Romania	ROM	Europe & Central Asia	Upper middle income
Russian Federation	RUS	Europe & Central Asia	Upper middle income
Samoa	WSM	East Asia & Pacific	Lower middle income
San Marino	SMR	..	High income: nonOECD
Saudi Arabia	SAU	..	High income: nonOECD
Senegal	SEN	Sub-Saharan Africa	Low income
Serbia	SRB	Europe & Central Asia	Upper middle income
Seychelles	SYC	Sub-Saharan Africa	Upper middle income
Sierra Leone	SLE	Sub-Saharan Africa	Low income
Singapore	SGP	..	High income: nonOECD
Slovak Republic	SVK	..	High income: OECD
Slovenia	SVN	..	High income: nonOECD
Somalia	SOM	Sub-Saharan Africa	Low income
South Africa	ZAF	Sub-Saharan Africa	Upper middle income
Spain	ESP	..	High income: OECD
Sri Lanka	LKA	South Asia	Lower middle income
Sudan	SDN	Sub-Saharan Africa	Lower middle income
Swaziland	SWZ	Sub-Saharan Africa	Lower middle income
Sweden	SWE	..	High income: OECD
Switzerland	CHE	..	High income: OECD
Syrian Arab Republic	SYR	Middle East & North Africa	Lower middle income
Tajikistan	TJK	Europe & Central Asia	Low income
Tanzania	TZA	Sub-Saharan Africa	Low income
Thailand	THA	East Asia & Pacific	Lower middle income
Tonga	TON	East Asia & Pacific	Lower middle income
Trinidad and Tobago	TTO	..	High income: nonOECD
Tunisia	TUN	Middle East & North Africa	Lower middle income
Turkey	TUR	Europe & Central Asia	Upper middle income
Turkmenistan	TKM	Europe & Central Asia	Lower middle income
Uganda	UGA	Sub-Saharan Africa	Low income
Ukraine	UKR	Europe & Central Asia	Lower middle income
United Arab Emirates	ARE	..	High income: nonOECD
United Kingdom	GBR	..	High income: OECD
United States	USA	..	High income: OECD
Uruguay	URY	Latin America & Caribbean	Upper middle income
Uzbekistan	UZB	Europe & Central Asia	Low income
Venezuela, RB	VEN	Latin America & Caribbean	Upper middle income
Vietnam	VNM	East Asia & Pacific	Low income
Yemen, Rep.	YEM	Middle East & North Africa	Low income
Zimbabwe	ZWE	Sub-Saharan Africa	Low income

Sectoral Growth Rates and ICOR under Different Sectors of the Economy

Sl. No.	Sector	Eighth plan		Ninth Plan		Tenth Plan	
		Growth Rate (%)	ICOR	Growth Rate (%)	ICOR	Growth Rate (%)*	ICOR
1	2	3	4	5	6	7	8
1.	Agriculture & Allied Sectors	4.69		2.06	4.05	3.97	1.99
2.	Mining & Quarrying	3.59	10.74	3.81	5.44	4.3	7.99
3.	Manufacturing	9.77	6.67	3.69	18.37	9.82	7.77
4.	Electricity, Gas & Water	5.5	18	6.46	15.43	7.99	14.97
5.	Construction	3.56	1.74	6.82	1	8.34	0.99
6.	Trade	9.06	0.54	5.86	1.09	9.44	0.91
7.	Rail Transport	1.95	27.94	4.7	9.87	5.4	14.66
8.	Other Transport	8.42	4.41	5.63	6.09	7.54	5.37
9.	Communications	14.31	7.25	17.14	5.28	15	8.33
10.	Financial Services	10.21	2.23	8.93	1.35	11.69	1.56
11.	Public Administration	3.91	7.82	9.21	4.09	6.43	5.45
12.	Other Services	6.22	4.19	8.19	3.7	9.26	3.53
	Total	**6.54**	**3.43**	**5.35**	**4.53**	**7.93**	**3.58**

Note: These are implicit ICORs calculated over the Plan period. For the Tenth Plan, these are targets. * Estimated
Source: Planning Commission, New Delhi.

Role of Agriculture in India

A flourishing agricultural sector is far more important for the development of Indian economy, since farming is less a business than a tradition in India. Even at this semi-industrialisation stage, about 80 per cent of its population still lives in the rural areas and directly or indirectly depends on agriculture for its livelihood.

The mere existence of about 85 crores of persons which are further multiplying at a rate of 2.5 per cent per annum depends on the developed agricultural sector. Further, most of our traditional industries on whose products depend our bulk of foreign trade and foreign earnings draw their raw material from this very sector. It is the surplus generated by this sector that would help Indian economy to reach the "Golden stage", since the total savings fund is made up of savings from the industrial sector and savings from the agricultural sector,

i.e., $I = S_i + S_a$

where, I = total saving fund.

S_i = savings from the industrial sector.

S_a = savings from the agricultural sector.

S_i is the function of profits generated by the industrial sector, which in turn depend upon the extent of demand created for industrial products by the agricultural sector. If the agricultural sector remains underdeveloped and fails to generate a matching demand for industrial products, profits will fall and we will soon be approaching a stage termed as "stationary state" by classical writers. Hence, the whole burden of increasing investment falls on the agricultural sector, so that.

$$S_a = TAS - f_a - i_a,$$

where, TAS is the total agricultural surplus

Fa is the farmer's consumption of agricultural goods

Ia is the farmer's consumption of industrial goods.

From above equation, *Sa* can be increased if we increase the total agricultural surplus, *TAS* and F_a and I_a could be maintained at the same old level. F_a and I_a can be maintained at the old level by the adoption of certain monetary and fiscal measures by the government. *TAS* on which depends our surplus and which in turn would initiate a process of development can be augmented only if we could increase the productivity of our agricultural sector.

We can, thus, say that in the absence of a developed agricultural sector, the base for "take-off" into a mature economy would be weak and Indian economy characterised by widespread disguised unemployment and a high rate of population growth, is expected to remain in a pitiable condition.

The role that the agricultural sector is playing in India at present can precisely be discussed under the following heads:

Share of Agriculture in National Income

The share of agriculture in national income is a crucial indicator of the role that agriculture plays in the economic development of a country. As the country rides on the wheels of progress, the relative contribution of agriculture in national income declines with the country becoming more and more prosperous. The expanding non-agricultural sector diverts surplus manpower from agriculture to industry and the improvements in agriculture enable a smaller number of people to produce for a larger population. With advanced agricultural technology, agricultural products are produced even for exports.

To the extent, therefore, the share of agriculture in national income declines, it marks a better level of economic advancement. On the other hand, agriculture is the single large contributor to national income. Therefore, a progressive agricultural sector means a higher level of national income and consequently, a higher level of economic development.

The Indian economy is still predominantly agricultural, about a half of the country's national income is derived from agricultural and allied activities which absorb nearly three-fourths of its working force.

Table: Share of Agriculture in Gross Domestic Product (At 1980-81 Prices)

Year	*National Income Rs. (crores)*	*Agricultural Income Rs.(crores) National Income*	*Agricultural Income as % of Total*
1970-71	90,426	35,930	39.7
1980-81	1,22,427	42,466	34.6
1990-91	2,12,253	60,991	28.7
1995-96	2,74,209	68,517	24.9

Two facts are being revealed. One, agriculture and allied industries contribute significantly a high share of the national income. Second, as hinted above, the share of agriculture in national income has been decreasing steadily.

The previous table clearly shows that agriculture even in recent years is by far the most important contributor to the national income, though under the impact of industrialisation, its share has been declining gradually. Its share which stood at 58.9 per cent in 1950-51 has declined to 39.8 per cent in 1983-84 while the share of industries which was 14.9 per cent in 1950-51 has risen to 21.8 per cent in the same period.

Supplier of Substantial Food and Fodder

The importance of the agricultural sector in India can be borne out from the fact that this sector supplies us the-necessities of life. Today, Indian agriculture is feeding about 100 million people, besides supplying other necessities of life. India's food production crossed the mark of 200 million tonnes in 1999-2000.

The agricultural sector is also providing all the fodder that is needed to sustain our livestock whose number runs into several crores. About one-fourth of the total world's cattle population live in India. The number of all sorts of livestock such as cattle population sheep, goats, horses, ponies, camels, etc. was estimated to be 45 crores in 1999-2000. The agricultural sector provides a variety of fodder to feed this large number of various types of animals.

Agriculture as a Source of Livelihood

Agriculture has a greater role in economic development in the less developed countries as it provides livelihood to a vast majority of people living in the country. This figure is not significant in terms of percentages only, but more so in terms of absolute numbers. The agricultural sector provides livelihood to about three-fourths of the Indian population, that is, seven out of every ten persons in India depend on agriculture. At the turn of the century, 71.5 per cent of the total labour force was engaged in agriculture and this situation has not changed until now. According to the 1991 Census, 69 per cent of the working force was still engaged in the primary sector. This indicates that in spite of rapid industrialisation in the country, the primary sector is still the main sector providing employment opportunities to the majority of the workers and has thus, acted as a big hock-absorber. This fact reflects the importance of agriculture and lesser development of other sectors of the economy.

Agriculture and Provision of Employment

Indian agriculture is of considerable importance in so far as it offers enormous scope of alternative employment. Past experience shows that the development of large industries has not helped significantly in absorbing the unemployed labour force. If has been estimated that since 1971, the labour force has increased by about 35 million. Of these, about 25 million have been absorbed in agriculture and 9 million in non-agricultural activities. Thus, the agricultural sector has absorbed a bulk of the additional labour force, though many must have got low intensity employment and consequently low incomes Agro-industries and agro-processing industries hold a great promise for employment in our country. The market has a close relationship with GDP at Factor Cost from Agriculture. The following table depicts important figures.

Gross Domestic Product(G.D.P.) at Factor Cost from Agriculture

(Rs. Crore)

Year	G.D.P. (Total)		G.D.P. (Agriculture)		Percentage Share of Agriculture	
	At Current Prices	At 1999-00 Prices	At Current Prices	At 1999-00 Prices	At Current Prices	At 1999-00 Prices
1	2	3	4	5	6	7
1999-2000	1786525	1786525	409660	409660	22.9	22.9
2000-01	1925017	1864300	408932	407176	21.2	21.8
2001-02	2097726	1972606	442464	433475	21.1	22.0
2002-03	2261415	2048287	425521	398206	18.8	19.4
2003-04	2538171	2222758	483030	441360	19.0	19.9
2004-05	2877706	2388384	501415	441183	17.4	18.5
2005-06	3275670	2612847	557118	468013	17.0	17.9
2006-07(Q)	3790063	2864309	634519	485937	16.7	17.0

(Q) : Quick Estimates

Source : Central Statistical Organisation, New Delhi.

Agricultural and Industrial Development

The role of agriculture in industrial development can hardly be over looked. In fact, agricultural progress is normally a prerequisite for industrial development. In a relatively closed economy, one of the most important preconditions of industrial expansion is the achievement of a rate of increase in agricultural productivity which exceeds the concurrent rate of increase in the demand for food.

Rising agricultural productivity supports and sustains industrial development in several important ways. First agriculture releases a part of its labour force for industrial development while meeting the increasing food needs of the non-agricultural sector.

Second, it raises agricultural incomes, thereby creating rural purchasing power needed to buy new industrial goods.

Finally, it enables agriculture to supply the major wage-good (food) of industrial workers at prices favourable to the profitability of new industry.

In fact, all growth models for surplus labour countries have underlined the importance of agriculture for developing the industrial sector in the initial stages. Agriculture supplies manpower and raw materials to industries and provides a wide market for industrial products. In India, most of our leading industries depend on agriculture for their raw materials. Cotton and jute textiles, sugar, vanaspati, and plantations all directly depend on agriculture.

There are a number of other industries whose dependence on agriculture is direct. These may include hand pounding and husking of rice, crushing of oil, weaving of handloom and khadi cloth, etc. Still other industries like paper, leather and tanning, matches, chemicals, etc. depend on allied activities of agriculture like forestry, animal husbandry, fisheries, etc.

It has been estimated that the industries which draw their raw material from the agricultural sector contribute nearly 50 per cent of income generated in the manufacturing sector in India.

Agriculture and International Trade

Indian agriculture has been a net earner of valuable foreign exchange for the country. Agricultural goods like tea, sugar, oilseeds, tobacco, spices, etc., constitute the main items of exports of India. Agricultural exports constitute a major portion of India's exports and accounted for 50 per cent of the exports at present.

This percentage will increase considerably if exports of jute manufactures and cotton textiles, whose 65 per cent of production cost is accounted for by raw jute and raw cotton, are included in agricultural exports. This has great significance for India's economic development because increased exports help the country to pay for increased imports of oil, machinery and raw materials.

Another important thing about agricultural exports is that while they earn a substantial share of foreign exchange, they do not drain it away through imports. As compared to manufactures, imports required for the agricultural sector have been much less. Besides, agriculture also contributes to national savings and capital formation.

We thus conclude that the agricultural sector occupies a central place in the national economy. The manner in which it contributes to the economic development of the Indian economy can be seen from the chart below:

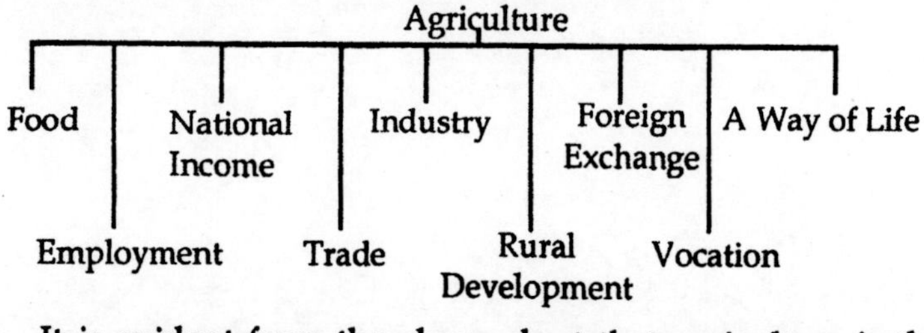

It is evident from the above chart that agriculture is the backbone of the Indian economy and development of this sector deserves to be accorded a very high priority in any scheme of resource-utilisation for general economic development. General economic development will require agricultural development either to proceed or to go hand in hand with it.

Indian Planners have already learnt a lesson during the second and third five year plan periods. It has been proved from the experience of Indian planning that failure of agricultural sector to deliver goods would spell disaster to the entire planning process.

3

Role of Agriculture in Economy

Land is the first thing that comes to mind when one talks of agriculture. In fact, it is not possible to think of agriculture without the use of land. Not that the use of land in non-agricultural pursuits is less significant but the nature of the industry of agriculture and its various processes make the use of land central to agricultural pursuits. The concept of land as used in economics is very complex and is held to stand for all natural resources. Land in agriculture may, however, be used in a restricted sense though its complex form affects agriculture in several ways. To be precise, our discussions will be pertinent to "farm land".

Land has been man's most significant input in his production enterprise and it continues to be so, more particularly in the less developed countries of the world. The size of landholding owned by a family unit determines its economic and social position in the society. Historically, man's affluence or poverty solely depended upon his control over farm land. Traditional methods of farming enabled him to increase farm output only by cultivating more land than he did before; as owing to the static technology, the possibilities of raising output through intensive cultivation did not exist. This resulted in man's everlasting demand for more and more land. Man's urge to possess more land has also been due to lack of alternative possibilities of employing his labour and capital. Such situations do exist even now in the poorer countries of the world where farming is the single most productive enterprise providing livelihood to three-fourths of the population. The desire for more land is also stimulated because of its permanent value, indestructibility and immovability. These qualities have tended to inflate the value of land and the prestige of its possessor.

Land is also held to be the safest form of saving and insurance against a future financial crisis. With the population rising fast, the demand for land for various uses shall ever rise and increase its value continuously. Investment in land has, therefore, all along made it a more prestigious asset than cash or other movable assets. In a similar fashion, changing technology has tended to increase land productivity considerably and made it an ever sought after asset in low-income countries. With demographic explosion on the one side and the poor employment capacity of the non-farm sector, on the other – the demand for land is bound to increase.

Role of Land in Agriculture

Land is the basis of agriculture. Virtually all types of production depend on land, in general, and all agricultural production depends on it, in particular. Besides being the resource for crop, fodder, and forest production, it also provides space for building cities, roads, airports, health resorts, schools, industries and so on. So far as the role of land in agriculture is concerned, it serves as the source of food and fibre production, pastures and grazing lands and forests.

Food and Fibre Production

Food production and the existence of mankind go together. Among the three basic needs recognised for the existence of man, viz., food, shelter and clothing, food is the most important of all. Land serves as the key resource in food production. A significant land area is, therefore, devoted to the production of food crops. Through the process of evolution, man learned to wear clothes and have shelter from rain and sun. Land plays no less a role in acting as the basic resource for fibre production and also in providing space for man to build his shelter.

Over time, not only has man's requirement for food and fibre increased, he has also sought different varieties of food/ Monocultures have since given way to multiple cultures. No longer do we raise a single crop from land, a variety of crops are raised to meet the multiple choices for man's food. Cropping patterns have to be adjusted in order to ensure better returns on the one side, and greater satisfaction, on the other. Changing technologies and tastes of people have led to frequent changes in the cropping patterns all over the world. Land has very patiently obliged man by offering itself to any form of exploitation that he designed. Ever growing requirements of food, fodder and fibre have led to the intensive use of land and this often reminds us of Ricardo's memorable phrase; "the original and indestructible powers of the soil". Land never felt exhausted nor shall it ever be to meet the food and fibre requirements of mankind.

Through technological development, of course, man has been able not only to restore to the land its inherent powers but has contributed a great way to improve these powers to raise higher levels of food output from a given stock of land. The importance of land as a basic resource of food and fibre production shall increase further as mankind's requirements will increase in future.

Pastures and Grazing Lands

Land is again a key resource for pastures and grazing lands which feed a growing livestock and help improve the supply of food to man. Since grazing lands do not fall under private ownership, the productivity of such lands is low. Nevertheless these lands play a significant role in feeding millions of cattle on the earth's planet. Most of these pastures and grazing lands are found in mountains where rugged topography and severe climate tell upon the productivity of land. The productivity of pastures and grazing lands can be improved considerably by promoting permanent pasture lands. It has been observed that permanent pasture lands under private control produce several times larger cattle feed than the unmanaged pasture lands. Keeping in view the grazing requirements of cattle feed, such steps as would increase the yield of grazing lands shall have to be taken and planned right now.

Role of Forests

Land under forests serves man in several ways, in general, and agriculture, in particular, Forest wealth plays a pivotal role in the development of a region and to agricultural enterprises.. Forests contribute to the development of a region in three important ways:

1. As a source of timber, firewood, medicinal herbs and aroma chemicals, minor forest produce and charcoal. There can also be the source of energy in terms of wind velocity and hydro dynamic pressure. Forest areas can also be developed as scenic tourist resorts. Forests are, in particular, important to agricultural development as these help in maintaining the ecological balance, soil moisture and perennial water supply.
2. Forests also offer scope for horticulture, plantation orchards, inter-culture, floriculture and pasture development and as such help directly and indirectly the development of agriculture in the region.
3. These also help in establishing agro-forestry-based industries and handicrafts.

As a source of timber and firewood, forests play an important role in the economy. Some forests are richly endowed with commercial timber which if exploited scientifically can contribute a lot to the development of the region in terns of income and employment. Forests also offer possibilities of intensive fodder development which may provide a basis for commercially viable animal husbandry programmes. Besides timber and firewood, forests supply a large number of minor products of immense utility.

A large number of agricultural enterprises can be undertaken on the forest lands such as horticulture, plantation crops, orchards, floriculture and pasture development. Hill lands offer ample scope for agro-forestry based industries and handicrafts. Forest lands provide a large variety of raw materials for setting up a number of small industries. These may include pulp, paper and hardboards, rosin and turpentine, sports goods, pencil industry, wicker work, match industry, drug and pharmaceuticals, aroma-chemicals, furniture and a large variety of other industrial enterprises. In a situation where farming refuses to absorb more workforce, what can be more ideal than to use these forest lands and their enormous potentialities to create alternative sources of employment?

The role of land in agriculture may also be examined by considering the functions of land as a factor of production in agriculture.

Land as a Factor of Production

Land provides space for any productive enterprise to take place and also acts as the repository of all the elements necessary for the growth of plants and animals. Providing floor space both for farming and factory is the most fundamental function that land performs. Since, however, farming requires a larger floor space compared to industry, a major chunk of land is devoted to agriculture even in the most industrialised countries. This is so because of the very nature of agricultural industry.

Secondly, land serves as a repository of physical, chemical and biological elements which are of utmost necessity for the" growth of plant and animal life. Unlike industry, agriculture greatly depends upon this function of land as without these qualities, land will become barren and unproductive. This while suggesting the significance of land in farming also suggests how important is the man's involvement in not exploiting the soil qualities alone but even in maintaining them at the desired level.

In fact, one of the definitions of agriculture is given as; "the utilisation of the natural fertility of the soil for the production of commodities serviceable to man". Man has an honest duty of making a genuine use of soil fertility and replenishing it before it is lost beyond repair.

The role of land in agriculture can hardly be overemphasised. Land shall continue to play a significant role in farm occupations and help in meeting man's requirements in this regard.

Non-agricultural Uses of Land

Although we are primarily concerned with the land in agriculture, a discussion on non-agricultural uses of land may not be quite out of place. The non-agricultural uses of land do have a bearing upon its efficiency in agricultural use.

The urban uses of land are enormous. With the growing urban population, the demand for land for purposes of urban settlement has gone up significantly. Though in towns, lesser lands are required to construct houses, build highways, hospitals, schools and public parks, yet the spurt in urban activities has been phenomenal resulting in a sharp rise in the demand for land in recent decades. Because of non-availability of employment opportunities in rural areas, developing countries are facing an unprecedented urban ward migration which has increased the demand for land for urban uses considerably. Greater industrialisation will exert a still stronger pull towards urban migration and land for urban uses shall be in great demand.

Land is required for recreation purposes. As a natural consequence of economic development and industrial tensions, the demand for recreation is increasing fast. It will be observed that recreation industry in the developed countries of the world is expanding faster than in the less developed countries where the extent of industrialisation is still low. Growing urban population and urge for outdoor visits have tended to decrease the per capital recreation land in the world. A large area of land is put under national parks, state parks and other public parks to provide open breathing space to the suffocated urbanites and industrial workers. Similar provisions in greater quantities shall have to be made for the future. Land shall have to be provided for fishing and hunting pursuits as well.

Another use of land is for purposes of transportation. Although it takes only a small proportion of total land, its use is fast increasing and in turn is affecting the other uses of land. Transport does not only claim a part of land but increases its use and efficiency for agricultural activities. Lands have to be provided for rail and road transport which is the single most vital item of infrastructure needed for speedier economic development. The land under highways and other roads is fast increasing and so is the case with the demand for land for the purpose of airports and railway lines. With increased mobility of man, the air, road and rail traffic continue to increase at a fast rate and more land shall have to be sought for these purposes.

Land also provides refuge to wildlife. The preservation of wildlife is set forth as one of the national goals and worldwide efforts are being made to save different species of wildlife from extinction. This is a known fact that clearing of larger forest areas encroached upon the liberty of flora and fauna and threatened its growth. Some of the wildlife is already extinct or some is on the brink of extinction and more areas of land shall, therefore, have to be enclosed for its preservation.

The use of land also arises for making provision for water reservoirs. Land is needed for the construction of dams, barrages and canals. As the demand for water both for purposes of drinking as well as irrigation is bound to increase, greater provisions of land shall have to be made for these purposes.

Finally, land have to be spared for mining production, sand dunes, bricks and other such uses as well.

Peculiarities of Land

Land as a factor of production has some important characteristics not common to the other factors of production. These characteristics may be listed as under:

1. Land is subject to the law of diminishing returns;
2. Land is limited in supply; and
3. Land is heterogeneous in quality.

Law of Diminishing Returns

This law is one of the oldest laws in economics. Its application in land is most pronounces. To a given piece of land, other things remaining the same, the application of additional units of labour and capital yield positive but diminishing returns for each additional unit of input. The law is founded on the vast and varied experience of farmers all over the world. There is a limit beyond which output per farm does not increase and this universal phenomenon is known to agriculture for centuries now. Additional units of labour and capital may give rise to higher returns for sometime but soon it happens that returns per unit of input become smaller and smaller till a stage is reached when an additional unit of labour and capital does not yield any increase in output. Beyond this stage, it will be judicious not to spend more on these inputs.

It is fairly easy to understand this phenomenon of land because if it were not so, the urge and necessity for more land would not have been there. Farmers always find it to their advantage to extend the area of cultivation rather than increase intensity of cultivation on a limited piece of land. By doing so, they succeed in reducing the incidence of the law of diminishing returns. This also explains the need for having larger areas under agriculture than under non-agricultural pursuits.

Limited Supply of Land

The second unique characteristic of land is that it is limited in supply by nature. In view of the fact that land is preeminently subject to the law of diminishing returns, the limitedness of its supply makes the problem more serious. Compared to other factors of production which can be expanded by man, land is not capable of expansion. This characteristic of land makes it a precious factor of production in agriculture. Not only are world's land resources limited by nature, a significantly larger land area is of no use in agriculture.

In India, for example, only about 45 per cent of land area is sown. In Great Britain not more than about 11 per cent of land is capable of cultivation. There are little possibilities of increasing the area under cultivation and for all practical purposes, land as a factor of production is fixed in supply.

Malthusian thinking on population and food supply was greatly influenced by the scarcity of land. He remained concerned with the uncontrolled growth of population, on the one hand, and fixed supply of land, on the other. Although the world has successfully avoided Malthusian dangers so far, the fact remains that unless the future size of population is regulated to the possible availabilities of food, mankind is bound to fall into the Malthusian trap.

Heterogeneity of Land Quality

The third important peculiarity of land as a factor of production in agriculture refers to its heterogeneity in quality. Although to avoid incidence of diminishing returns, man has ever sought to bring more land under the plough, he has always faced problems on account of the poor quality of new lands. Naturally, the best lands come under the plough the first and lands of poorer quality are sought thereafter. Not only does heterogeneity arise as such, different types of soil qualities, ecological conditions, topography and other factors also lead to wide varieties in the quality of land. It is these variations that give rise to economic rent.

Possibilities of Expanding Land Area

The fact that land is limited in supply is universally recognised. What then could be the possibilities of expanding land area for agricultural and non-agricultural uses? Since the world population is rising at a fast rate, the pressure on the existing land resources is mounting and adverse man-land ratios result in declining marginal productivity of labour and other inputs. Under such circumstances, poorer quality lands have to be brought under cultivation.

Though these lands tend to show signs of poverty, there probably would be no escape from extending the agricultural operations to these areas. Fortunately, technological development has resulted not only in maintaining the soil qualities of land for longer durations but also in improving the quality of poorer lands. Such extensions, whatsoever their problems may be, are also subject to a limit.

Perhaps the only possibility of expanding land area is by reclamation from deserts, forests and river-ways. Such possibilities may have appeared to be enormous in the beginning but eventually when the process begins, it no longer looks like that. Obviously, such reclaimed lands will be of poorer quality, but the same have to be reclaimed to reduce pressure on the existing land resources.

As already mentioned, the reclaimed poorer quality lands can be improved qualitatively by using modern technology. Such improved technology operates in different forms. By disseminating knowledge and research findings, the system of farming may be improved to yield better results. Farm management may become more scientific and help organising low quality farms in such a way as to reap a better harvest than would ordinarily be expected. However, there have been apprehensions that modern technology is likely to widen the gap between good quality land and reclaimed poor quality land. Some of the recent works done on American agriculture have shown that modern technology has tended to harden the upper crust of land and the depth of hard crust is increasing. This further weakens the moisture retention power of the soil and impoverishes the soil quality.

Under such a situation, Americans have always been feeling compelled to use bigger tractors that can dig deeper. This process is, however, increasing the gravity of the problem rather than solving it. If this is happening to the currently productive lands, what then will happen to the poor quality reclaimed lands is not hard to guess. However, the possibilities of further improvements in modern technology will always be there to subdue its negative effects.

Modern technology may also come in a big way in the development of land for agricultural purposes at a reduced cost. Tractor technology may be cited in this regard. Levelling of rugged surfaces and building roads have become not only less expensive but also timesaving. Traditionally labour-using devices have been used partly because modern technology was not known and partly because opportunity cost of labour was low. If heavy machinery has to be diverted to land development, it certainly would be at a much higher opportunity cost and to justify such a diversion, the returns from land so developed should be sufficiently high.

Reclaimed low quality lands can also be used for putting up new colonies and settlements if these areas are fully developed to suit human habitation. Modern technology can again come in a big way in this regard. Instances could be quoted from Sri Lanka and South Vietnam where modern technology helped eradication of malaria from some areas resulting in their becoming fit for human settlement. In this way, a lot of pressure on existing land resources for non-agricultural uses can be reduced significantly.

While formulating land reclamation plans, however, due consideration shall have to be shown to the cost component. Obviously, if cost involved is greater than the returns expected, such a reclamation policy may not be favoured. The important cost components are labour and capital while as in some developing countries, labour may be available in large quantities and relatively at a cheaper rate, but capital is highly scarce and cannot easily be diverted from its current use. The opportunity cost of labour may be low, as in surplus labour countries, a lot of labour force is either idle or engaged in low-income callings; the opportunity cost of capital, however, is very high and diverting it from its current use will certainly involve heavy costs. The high cost components on the one side and the dire necessity of reclaiming new lands, on the other, is bound to create a serious dilemma for the future planners. The cost component shall, however, have to be balanced with the return components before plans for land reclamation are put through. Whatever the case may be, the fact remains that land area shall have to be expanded through reclamation and plans for maximising returns from such lands shall also have to be formulated. One thing that may be noted is that reclaimed lands may create supplementary avenues of reaping benefits (such as creating more marketable agricultural surplus, earning more foreign exchange and widening the national market) which may more than proportionally compensate the costs involved in such a project.

It is also possible that some good quality lands may still be available in the less populated areas of the world as Africa and Latin America. Lands may also be released for productive purposes from such agencies as are not putting these to proper use. Some institutional changes may have to be brought about in this regard.

4

Modern Indian Agriculture

The agriculture sector continues to be the backbone of Indian economy contributing approximately 27.4 per cent to the gross domestic product (GDP), and accounts for about 18 per cent share of total value of country's export. The agricultural production has kept pace with the popular growth rate of 21 per cent per annum. Today we are second largest producer of wheat, rice, fruits, vegetables, and fresh water aquaculture; and largest exporter of spices and cashew.

The late sixties and seventies were the years of Green Revolution. During Yellow Revolution oilseeds production reached up to 24.4 million tonnes. Per capita availability of foodgrains went up to 528.77 g per day in 1996-97 when compared to 395 g in early fifties. Fertilizers consumption has also increased and India has become fourth in the world after USA, USSR and China. Pulse crops are grown on the largest Indian area in the world and India is the first to evolve a cotton hybrid.

Cropping pattern is changing and commercial crops and non-traditional (moong, soyabean, summer groundnut, sunflower, etc.) are gradually growing more importance in line with domestic demands and export requirements. Short duration varieties have been introduced to use the residual moisture available from post-kharif and post-rabi cultivation.

The index of agricultural production base T E 1981-82 = 100, recorded following trend

A fall of 2%	:	1991-92
An increase of 4. 1 %	:	1992-93
An increase of 3.8%	:	1993-94
An increase of 4.9%	:	1994-95
A fall of 0.4%	:	1995-96

Foodgrain production was quite low after Independence because high-yielding area of Punjab went to Pakistan after division of India. In 1950-51 the food grain production was 51 million tonnes but it was 193.01 million tonnes during 1999-2000 resulting in a buffer stock of 35 million tonnes.

Land Use

Land utilisation statistics revealed that net sown area increased from 1,187.5 lakh in 1950-51 to 1,424.2 ha in 1998-99. The relative share of foodgrains and non-foodgrains in gross increased from 404.8 lakh ha to 682.8 lakh ha in the same duration.

Harvesting

The 3 main crop seasons are - kharif, rabi and jayad. Major crops are rice, jowar, bajra, maize, cotton, sesame, soyabean and groundnut. Major rabi crops are wheat, jowar, barley, gram, linseed, rapeseed and mustard. Rice, maize and groundnut are grown in the summer also.

Three types of seeds, namely, breeder, foundation and certified, are recognised by the system. Indian seed programme include central and state ICAR, SAU system, Public sector, cooperate sector and private sector institutions. National Seeds Corporation (NSC), State Farms Corporation of India (SFCI), 13 state seed corporation (SSC) and about 100 major private sector seed companies are the main components of Indian Seed, State Seed Certification Agencies (SSCAs) and 19 State Seed Testing Laboratories (SSTLs) looks after quality control and certification. The Seeds Act, 1966, provides:

1. Legislative framework for regulation of quality of seeds sold in the country.
2. System of certification of seeds sold in India.
3. Notification of varieties, a prerequisite certification Administration of the Act and quality control of seeds to look after by control solved committee and its various subcommittees and Central Seed Certification Board.

Seed has been declared an essential commodity under the Essential Commodities Act, 1955. The Seed (Control) Order 1983, was promulgated to control and regulate the seed production and distribution. The new seed policy on seed development has been in operation since 1988. The main aim of seed policy is to makes available the best quality seed planting material "to farmer" anywhere in the world. Plants, fruits and seeds (Regulation of import into India) Order, 1989, regulates the plant quarantine clearance. Export of seeds is liberally allowed, only certain categories of seeds and planting material are in list of restricted climate for which a licence is required.

National Seeds Project III (NSP III) aimed at overall important in seed quality programme. Since 1969, the Central Seed Committee has verified 2,385 varieties of agricultural and horticultural crops.

The consumption of chemical fertilizer during 1999-2000 is estimated to be more than 14.93 million tonnes. Sharp increase in prices and introduction of biofertilizers result in its lower consumption.

The Government of India is implementing two Sponsored schemes: (i) balanced and use of fertilizers to popularise the use of organic sources of nutrients (compost, green manure, biofertilizers; etc., and (ii) National Project and Technology Mission on Development and Use of Biofertilizers - to give adequate thrust to biofertilizer production and promotion under Essential Commodities Act, 1955. The government has issued Fertilizer Control Order, 1985. The government has been implementing a Central Sector Scheme on strengthening of Central Fertilizer Quality Control and Training Institute since Fourth Plan.

Water and Soil Conservation

Soil and water conservation measures were launched in First Five-Year Plan. Till the end of 1995-96, 15.22 per cent of the treatable area had been treated in the area catchment of River Valley Project. Under Flood Prone River Scheme 10.25 per cent area of the total treatable area had been treated till the end of 1995-96. Under the Seventh Five-Year Plan, a centrally-sponsored scheme of reclamation of alkali usar soils was initiated in Haryana, Punjab, and Uttar Pradesh. It was extended to Gujarat, Madhya Pradesh and Rajasthan. During 8th Plan Watershed Development Project in Shifting Cultivation Areas (WDPSCA) has been initiated in northeastern states. It was in accordance with the guideline of the on going centrally scheme of National Watershed Development Project Rainfed Area (NWDPRA).

Machinery and Implements

Farmers have been provided assistance for owning agricultural machinery including tractors. Besides this farm machines are exhausted for their characteristics and betterment. Five states agricultural universities are being aided for farm machinery testing, training and human resources development. In spite of the efforts the improvement in farm machinery use has been mainly northern states and in few areas where irrigation facilities have been developed.

Sale of tractors (220.937) and power-tillers (11,000) touched the all time high work in 1996-97, and because of this at farm available power in 1.10 hp/ha in 1996 compared to 0.35 hp/ha in early 70's. During the Ninth Plan, the main emphasis was on improving and popularising animal/power driven implements and small farms. Water saving devices like sprinkles and drip irrigation were given main importance. During Eighth plan, centrally-sponsored scheme, promotion of agricultural mechanisation, small farmers was launched and under it 30 per cent subsidy limited to Rs. 30,000 was given to farmers, then groups, etc.

During Ninth plan two schemes, viz. (a) promoting/ popularisation of agricultural equipments in north- eastern states, (b) conducting studies and formulating long-term mechanisation strategies for each agroclimatic zone, were started. State Agro-Industries Corporation (SAICSs) act as catalyst for providing to the farmers access to various industrial inputs for agriculture.

Power threshers has been brought under Dangerous Machines (Regulation) Act because of increasing awareness on safety measure among users.

Protection of Plants

Integrated Pest Management (IPM), in eco-friendly approach, was adopted as a major thrust area of crop protection during 8th plan. IPM aims at minimal use of hazardous chemical pesticides by using alternate pest control methods and technique. The surveillance for pests and diseases on economically important crop was carried out and about 1,603 million bio-agents were released. Bacillus neem-based bio-pesticides are being granted regular registration status under the Insecticides Act, 1968. BHC, Aldirin, cheorclane, Heptachlor were banned with effect from April 1997 and 20 September, 1996, respectively.

The plant guarantee activities aim at presenting introduction of exotic pests and diseases into the country by adopting domestic plant qurantine regulation. It also works for controlling / containing of those exotic diseases/pests already introduced in the country. Destructive Insecticides and Pests Act (DIP Act) and Plants, Fruits and Seeds (Regulation of Import into India) Order, 1989 (PFS order) are implanted under this scheme.

Phytosanitary certificate (PC) as per provisions of International Protection Convention (IPC), 1951 of FAO, is also supplied under these activities. The post-entry quarantine stations located at 10 international airports, seaports, and 7 land frontiers. National Plant Protection Training Institute, Hyderabad, provides training in various airports of Plant Protection.

Total net sown area in the country is about 142 million ha and out of it 92.6 million ha is rainfed. Production has wide fluctuation due to variation in rainfall as whole farming depends upon rainfall which is often erratic and unpredictable. Jowar, bajra other millets pulses, oilseeds and cotton are grown. Increase and stability in their productivity is of crucial importance. In the 8th plan holistic approach was adopted under the National Watershed Development Project for Rainfed Area (NWDPRA). It aimed at development of integrated farming system on watershed basin in shifted areas. The NWDPRA was launched in 1990-91, with the objective of restoration of ecological balances in rainfed areas and sustainable bio-mass production. It also works for generating employment opportunities for rural manner in rainfed areas by developing self-help groups. Under this project the target is of treating 28 million ha area. The Integrated Watershed Development Project (Hills) and Integrated Watershed Development Project (Plains) are underway and their main aim is slow down and reverse degradation of natural environment and improve potential of the areas.

Agricultural Development Projects (ADPs) with the assistance of World Bank are being implemented to (i) enhance long-time sustainability in agriculture, and (ii) create infrastructure in rural area.

DANIDAG (Government of Denmark), EEC (European Economic Community) and SDC (Swiss Development Corporation) are playing vital role in aided agricultural development programmes.

The Technology Mission on oilseeds provided the necessary breakthrough in oilseeds production. Oilseeds sector has now become to major foreign exchange earner. The oilseed production was doubled in decade (12.6 mt in 1987-88 to 24.4 mt 1996-97). Time is popularly famous as Yellow Revolution. Pulses and maize were also brought under the ambit of Technology Mission. Centrally-sponsored Oilseeds Production Programme (OPP), National Pulses Development Project (NPDP), and Accelerated Maize Development Programme (AMDP) are being continued in 9th Plan also.

The production of fruits and vegetables was 38.3 and 67.24 million tonnes respectively. India attained second position in coconut production. Above 10 per cent of coconut production is from Kerala, Tamil Nadu, Andhra Pradesh, and Karnataka. India is the world's, leading producer and exporter of cashew accounting for nearly world's 50 per cent demand.

Animal Husbandry

The gross value of output from livestock sector, at current price, was about Rs. 827 billion in 1995-96 about 26 per cent of value of total output agricultural sector excluding animal draught buffaloes, 50.8 million sheep, 115.3 million goats, 12.8 million pigs, 3 million pack animals and 307 million poultry.

The Operation Flood Programme, the world's, largest integrated dairy development programmes, completed its III phase in 1996. By September 1990, about 73,300 dairy cooperatives were organised including about 9.4 million farmers. The average milk procurement in July 1997 was

107.3 lakh litre milk/day and average milk marketed per day was 112 lakh litre milk per day. More than 62 per cent milk procurement was from marginal, small and landless farmers. The milk production was almost stagnant between 1947 and 1978 with an annual growth rate of merely 1 per cent, but after that Operation Flood Programme a growth rate 4.5 per cent per annum was observed. In 1996-97, the milk production was 69 million ml tonnes.

In 1969-70 the milk availability per capita was only 107 ml/day which increased to 202 ml/day in 1996-97, which is still under the recommended nutritional requirement of 220 ml/day as per Indian Council of Medical Research.

The Government of India launched Technology Mission on Dairy Development (TMDD) in August 1988 to accelerate the pace of Dairy Development in country and reached to first position in the world. Father of White Revolution is Prof. (Dr.) V. Kurien.

Blue Revolution by enhancing fish production from 0.75 million mt in 1951 to 5.4 million mt in 1997 put India as second largest fish producing country in the world in fresh water aquaculture. Fish production increased at an average rate of 4.4 per cent 1 year during the five years of 8th Five-Year Plan. Fish Farmers' Development Agencies (FFDAs), an important programme, was launched by the government to improve overall production of inland fisheries in India. These agencies brought about 3.87 lakh ha water area under intensive fish culture. There are 6 major fishing harbours, viz. Cochin, Chennai, Vishakhapatnam, Roychowk and Paradip, and 41 minor fishing harbours and fish landing centres. Brackishwater Fish Fanners. Development agencies are trying to improve the countries' vast brackishwater area for shrimp culture. Guidelines has also been issued for mitigating the adverse impact if any, of farming on the coastal ecosystem. The government of India started two important schemes for the welfare of traditional fishermen. These are - (i) Group

Accident Insurance Scheme, and (ii) Development of Model Fisherman Villages.

Census of Agriculture

As a part of the World United Nations Agricultural Census Programme in pursuance of the recommendation of the FAO, the Department of Agriculture and Cooperation has organised census on quinquennial basis since 1970-71 (1970-71, 1980-81, 1985-86, 1991-92, 2000-01). Sixth agricultural census with reference 1995-96 is in operation. According to report of 1990-91, operational holdings in the country has increased from 972 million in 1985-86 to 1,066 million in 1990-91 and operated area has risen by 0.6 per cent.

The press of population on land with average size sure of holding decreased from 1.69 ha in 1985-86 to 1.55 ha in 1990-91. Holdings operated by scheduled castes has increased by 11.5 per cent and by scheduled tribes by 4.2 per cent from 1985-86 to 1990-91. In 1901 the population in rural area was 89.2 per cent while in 1991 it reduced 74.3 per cent. In 1991 the total number of villages was 580,781 out which 1,12,803 are in Uttar Pradesh only.

National Bank for Agriculture and Rural Development (NABARD) was launched on 12 July, 1982. It was established for promotion of agriculture, rural-side industries, village industries, handicrafts and other activities in rural areas to promote rural development.

Landless Agriculture Labourers Group Insurance (LALGI) Scheme is in operation since 1987. It takes care of heads of families in the age group of 18-60 years and who are not land holders. Under IRDP a Group Life Insurance Scheme is being separated for which entire premium is paid by the central government.

Programmes for Houses

The average duration of farm and home broadcast 60-100 min. Farm schools as method of communicating distant education on fanning have been adopted by All India Radio stations located in different regions. Akashvani Annual Award is given away to best Farm and Home Programmes of the year.

The Directorate of Extension (DoE) undertake agricultural extension programme of the Government of India.

The National Institute of Agricultural Extension Management (MANAGE) at Hyderabad, looks after the human resource development activities of Directorate of Extension (DoE). DANIDA is helping agricultural training.

Educational Research

The Indian Council of Agricultural Research is directly involved in undertaking fundamental as well as applied research in traditional and frontier areas to offer solution to problems relating to agriculture. The research is carried out through a chain of 45 institutes and 4 national bureaux, 30 national research centres, 10 project directorates, 80 All India Coordinated Research Projects, 4 National Institutes and Deemed-to-be university status, 31 State Agricultural Universities and 200 zonal research stations.

The whole country has been divided into 120 district agroclimatic zones. Large network of front-line ICAR-SAU extension system consists of 261 Krishi Vigyan Kendras, 8 Trainers., Training centres (TTC) and 42 Institutions Village-Linkage Programme (IVLP). The IVLP was started for technology assessment and refinement. The ICAR has established Agricultural Information System (ARIS). A WAN connecting SAUs, ICAR institutes/headquarters has been established. Connectivity is provided through dial-up, leased line, RF links and VSAT (100). It allows E-mail, Internet browsing-file transfer, etc.

The NATP was jointly prepared by the ICAR and Department of Agricultural Cooperation (DOAC), and for this the World Bank assured assistance of Rs 200 million. Its main aim is technology generation and its transfer. Its three major components are - Agro-ecosystem Research, innovations in technology dissemination, organisation and management system.

New Trends

The perspective plans Vision 2020 formulated up to 2020 AD for sustainable growth of Indian agriculture. The required per cent growth in production of various commodities over the 1994-96 average to meet the demand of by 2020 and is very high.

Items	*Growth Rate % over the 1994-96*
Foodgrains	55
Fruits	142
Milk	28
Meat	57

5

Indian Agricultural Policy

Indian Economy after passing through several phases of development has come to pass a very difficult phase towards the beginning of nineties. These unpleasant developments on the economic front had emerged out of both domestic and external disturbances primarily from political instability which was being witnessed both on home as well as an international front. Foreign exchange position had primarily come under heavy strain.

The liquidity crisis of 1991 turned out to be a watershed in India's post-independence history. The balance of payments situation had deteriorated so sharply and the foreign exchange reserves had fallen so low that the possibility of default in payment was imminent. Orthodox and unorthodox measures had to be taken urgently to restore credibility. On the domestic side while the Indian economy had done extremely well in terms of real growth between 1985 and 1990, the fiscal situation had deteriorated sharply.

The budget deficit as well as the overall fiscal deficit had sharply increased contributing, on the one hand to large increases in money supply and, on the other, to sharp increases in interest payments. Fiscal deficit of the centre and the states taken together which was about 7.5 per cent of the GDP in the late 70s had increased to about 11 per cent by 1991. The fiscal deficit of the central Government alone which was between 6.00 per cent in the late 70's had increased to 8.5 per cent during the same period. Consequently, interest payments in the central government's budget had become the single largest expenditure item rising from 2 per cent of GDP in 1980-81 to near 4 per cent of GDP in 1990-91. The country thus entered the 90s with a fiscal deficit that was not sustainable. These disturbing developments led to a review of economic policy and forced unavoidable changes in it.

The New Economic Policy comprises the various policy measures and changes introduced since July 1991. There is a common thread running through all these measures. The objective is simple and that is to improve the efficiency of the system. The regulatory mechanism involving multitudes of controls had reduced competition even in the private sectors. The thrust of New Economic Policy was towards creating a more competition environment in the economy as a means of improving the productivity and efficiency of the system. This was to be achieved by removing the barriers to entry and the restrictions on the growth of firms. What was sought to be achieved was an improvement in the functioning of the various entities, whether they be in the private or in the public sector, by injecting an element of competition.

The policy changes brought into force since 1991 fall broadly into two categories. The first set of measures is part of what is normally known as stabilisation policy. The second act of measures come under the category of structural reform policies of while the stabilisation policies are intended to correct the lapses and put the house in order in the short-term, the structural reform policies were intended to accelerate economic growth over the medium term. Structural reform policies cannot succeed unless a degree of stabilisation has been brought about. But stabilisation by itself will not be adequate unless structural reforms are undertaken to avoid the recurrence of the problems faced in the recent period.

Structural reforms were broadly in the area of industrial licensing and regulation, foreign trade and investment and financial sector. There is considerable unanimity among the economists about the need to reduce and as far as possible eliminate barriers to entry and expansion of firms. The policy of licensing as has been practised in the past had no particular merit. In relation to foreign trade policy, the aim was to liberalise the regime with respect to imports and try to bring about a closer link between exports and imports. Yet another objective is to reduce the tariff rates. As regard foreign investment, the new policy measures certainly make a break with the past.

In an era in which capital is mobile and moving across boarders in a big way and where technology transfer is through investment, we cannot afford to close our country to the flow of foreign investment. Finally in relation to the financial sector, it has to be noted that while there has been a considerable widening and deepening of the Indian financial system many inefficiencies have crept into the system during the past 20 years. A reform of the financial system to provide greater autonomy to the institutions both in terms of interest rate structure and operational matters had become necessary.

Content of Reforms Programme

The programme recommends in its extreme form full economic liberalisation which implies: *(a)* elimination of industrial licensing, capacity controls and capital issue restraints; *(b)* relaxation of MRTP regulations on the growth of big business houses; *(c)* removal of administrative control in prices and distribution of commodities, public utilities and labour markets; *(d)* reforms of credit and financial system in a manner as would allow the banking system to play its true role as a financial intermediary; *(e)* fiscal reforms aiming at elimination of deficit financing and inflation and at introducing tax reforms so as to reduce the size of tax evasion; *(f)* price-reforms for products of public and enterprises; *(g)* closure of sick units in the public and private sectors and at least partial privatisation of public units to make them financially viable;

(h) import liberalisation by elimination of quantitative restrictions and replacing them by tariffs; *(i)* reforming tariff structure by reducing the average rates of tariffs; *(j)* liberalising imported inputs for exports and adopting flexible exchange rate policy; *(k)* improving the availability of foreign exchange by opening up domestic capital markets to nonresidents and foreign institutional investors; *(l)* improving the availability of foreign exchange through direct investment flows by lowering entry barriers on them; *(m)* improving country's access to international capital markets through external debt reforms and finally; *(n)* introduction of current and capital account convertibility.

With the adoption of new economic policy, there has been an increasing trend of withdrawing of controls and regulations in the economic field. A new economic environment has, thus, come into being which calls for a new approach and strategy to deal with the situation. An impression is usually carried that the reforms of the type as mentioned in new economic policy have a bearing only on industry and infrastructure. This is, however, not so. Agriculture cannot escape the important policy changes; whatever these may be and whenever these may occur. Unlike industries, agriculture in India has functioned more or less in a competitive environment all along.

Nevertheless, "inefficiencies in resource use in agriculture arose due to trade restrictions, inappropriate pricing of inputs and outputs and the prevailing macro-environment. Trade restrictions take various forms like, total ban on export, export licensing, Canalisation through Government agencies, as well as restrictions on export prices. Distortions in input prices arise mainly due to subsidies on farm inputs. A combination of the trade restrictions and Government intervention in commodity markets result in distortions in farm output prices. It is quite important to note that the agricultural and rural sector which form the ministry of two-thirds of the population in India are also the focus of attention now for significant policy changes and private initiatives as part of the new economic regime. Already steps like reduction in subsidies and removal of some restrictions on domestic and international trade in agricultural commodities have been and are being taken.

All this is being attempted with a view that agricultural sector can really act as the most effective safety net in the process of adjustment programmes both in terms of giving positive results as well as preventing some of the negative effects of a market economy. Impact of new economic policy on agriculture and agricultural development has been discussed here under five heads: (i) fiscal adjustment; (ii) the structural adjustment policy; (iii) financial reforms, (iv) policy of globalisation and (v) other measures and approaches relevant to the agricultural sector.

Fiscal Adjustment

Since July, 1991, a prime component of the Government's drive to contain the fiscal deficit within limits has been the reduction in food and fertilizer subsidies. However, there has been no reduction in budget allocation for food subsidies so far, the budget for 1998-99 provided Rs. 9000 crores as against the 1992-93 revised estimates of Rs. 2800 crores. Although the level of subsidy in the Indian agriculture sector estimated at about 5 per cent of the agricultural product is much below the permissible limit of 10 per cent under GATT provisions, yet the budgetary burden of agricultural inputs and output subsidies is becoming progressively. Unbearable for the national exchequer.

An immediate impact of new economic policy on agriculture could, therefore, be reduced budgetary allocations for this sector. This would call for phasic, if not drastic reduction in agricultural subsidies to begin with. Subsidies for agricultural inputs may warrant reduction and so may be the case as far as support to the output prices is concerned. This would directly affect cost of production, returns and profit margins in agriculture sector, specially for foodgrains crops, because these are the crops that are major users of subsidised inputs and derive the main benefit on account of support and procurement prices. Again, reduction in tariffs on imports of competing products, allowing the imports more freely may pose stiff competition for the farmers in the domestic market in the years to come.

Thus in a situation of resource crunch, which India is facing today, the agricultural sector will have to perforce more away from the regime of subsidies. "Easy situation on foodgrain stocks is likely to impel the government to go easy on administered prices and procurement through support price system may also not remain in demand to that extent. Unless these surpluses find reasonable export market, in a competitive market environment, there would develop pressures for diversification of production patterns away from foodgrains and traditional crop systems to high value products that would be suitable for processing and would be more in demand in the national and international markets.

Indian agriculture, thus operating in a free competitive market, will have to face international competition in both the domestic and international market". Many aspects of macroeconomic stabilisation programme may result in fiscal compression resulting in withdrawing subsidies to fertilizer, deceleration of public investment in irrigation, power and rural and infrastructure including agricultural-research roads and communication, etc.

Structural Adjustment Policy

The basic logic of the SAP is that industrialisation will set in motion forces which generate commensurate growth impulses in other sectors of the economy as well. The condition which the SAP introduces for this kind of industrialisation and growth are: (i) substitution of the market and of private enterprise for planning and public sector in industry; (ii) re-orientation towards export production in phase of import substitution, and (iii) removing the capital goods industries bias in resource allocation and letting the market do the allocation. As a consequence of the above conditions for industrialisation and growth there would be downgrading of the priority to food self-sufficiency and emphasis would be placed on more commercialised and export-oriented rural and agricultural economy. In order to achieve high agricultural productivity as a part of NEP philosophy, a substantial increase in the average size of the farming which has to be combined with considerably more capital intensive technologies.

As agriculture becomes more market oriented. "Government Policy regarding procurement and maintenance of food stocks would also change – Except for such restricted public distribution outlets as may be maintained for purely political reasons, buffer stocks may be held only for occasional market intervention by government. But even this may become redundant when the Indian market is globalised, since imports may take over the job of filling in the shortfalls in domestic production when necessary." However, in persuing above economic programme, we may not loose sight of the costs of transition in terms of agricultural unemployment and other sacrifices to be made by marginal and small farmers. Above improvements in agriculture doubtless, require less labour and more capital investment.

The new economic policy shall have to be weighed against Indian agricultural situation keeping in view the national priority of food self sufficiency and accelerated agricultural transformation so as to generate more surpluses not only for export but also for industry. Keeping also in view that trickle down mechanism is not effective, more so when greater emphasis is on privatisation, any distortion in agricultural development shall seriously handicap the working of the welfare state towards removal of poverty and unemployment. Notwithstanding the desirability of reforms, influences on agricultural and rural sector will have to be studied carefully and non-conflicting safeguards provided for to ensure harmonious growth of the economy.

Financial Reforms

The rural credit svstem is threatened by rising overdues and deficiencies in management. Improving the economic viability of credit institutions, is, therefore, high on the agenda for reforms. The high priority accorded to it is clear from the fact that within months of its assuming office. Narasimha Rao government appointed the Narasimha Rao Committee on the Financial system. Though committee has confined its attention to the industry and export sectors, yet it takes the view that the rural branches of commercial banks are unremunerative and that the Government's policy of requiring banks to provide 40 per cent of their advances to priority sectors has been a drag on their resources.

Such a recommendation of the committee has resulted in something worse than a being neglect of the agricultural sector. R.B.I. has permitted banks to close down loss-making rural branches, and there is no longer any urge or incentive to extend the rural network. However, recently, Government has augmented the share capital of the National Bank for Agriculture and Rural Development (NABARD) and undertaken a major programme for rehabilitating the financially weakned Regional Rural Banks. It has also deregulated the leading and deposit rates of all credit rate of 12 per cent per annum in order to give cooperatives greater flexibility to mobilise resources and lend them to their members at appropriate interest rates which cover costs.

Globalisation and Agriculture

The working of new economic policy will force the Indian agriculture to drop the crutches of support procurement-administered prices and farm subsidies mechanism and "also out of its own necessity and in response to GATT challenges and opportunities that will emerge in international markets in the near future, agricultural market is bound to get progressively integrated with the global market." The Government of India's drive towards globalisation has been motivated by three principal expectations: *(i)* the entry of foreign competitors into India's domestic market will stimulate local producers to minimise their costs of production; *(ii)* with reduced cost and access to imports India's exports can be rapidly expanded; *(iii)* with freedom of access to foreign investment the economy will benefit from an inflow of both financial resources and advanced technology. These factors are, also expected to transform Indian agriculture and elevate its growth rate.

However, it has been demonstrated by many studies that agricultural production is more responsive to non-price factors like irrigation, technology, credit, infrastructure and so on. Though trade liberalisation in terms of reduction in imports restrictions and tariffs and export taxes can help reduce the anti-export bias of the previous policy but this does not necessarily mean an increase in the Real Exchange Rate (RER) as the other macroeconomic policy changes in the fiscal financial and monetary sectors tend to offset the increase in RER resulting from trade liberalisation measures after the expansion in production for export.

Enlarging agricultural exports depends on a continuing increase in agricultural output, together with the diversion of a larger proportion of it to the export sector. Under free market system, this implies the following:

1. a high price-elasticity of agricultural production;
2. existence of higher prices abroad than the domestically prevalent;
3. availability of adequate shipping and other facilities to shift supplier from one market to another; and
4. easy access to necessary finance and market information.

Unfortunately, all these conditions do not exist presently. And if they do not, can they be created quickly enough to push up exports and export earnings? In brief the export potential for each agricultural commodity should be reviewed separately. Much depends on the efficiency and productivity. It requires vigorous implementation of the measures to improve the productivity by upgrading technology and to evolve a strategy for global competitiveness.

This has to be a coordinated effort involving government-cooperatives, research organisations and farmers. Even though India has many agricultural Universities and Research institutions, technological upgradation will be an uphill task and effective extension on a big scale. Can Indian agriculture be an equal partner in the global system, is an important question. However, bigger and broader exposure through liberalised regime should open out new opportunities and stimulate innovations in production as also in marketing of agricultural products in the international market.

Multinational Companies

As a consequence of liberalisation and globalisation of the Indian economy, it is but expected that Multinationals will enter the domestic market. The activities of these corporations have a significant impact, positive or negative on the economic structure and political and social systems of the countries where they operate.

The MNCs are interested in capturing markets in host countries because the domestic demand may be inadequate. MNCs have played an innovating and catalytic role in funding new industries, transmitting technological and managerial skills as well as capital and, in many cases, creating entire social infrastructures, including transportation, in order to conduct their business.

Moreover, MNCs have world wide marketing organisations that facilitate exports from developing countries and thus help the process of transformation of a traditional less productive sector into high productive export sector. "Opportunities that emerge, through the entry of MNCs are tremendous by way of forcing the domestic companies to become quality conscious and cost effective both in the supply market and product market.

Also the induction of new technology and quality parameters in collaborative enterprises can go a long way in improving production and value added processing of agricultural produce. As the process catch up, more avenues of gainful employment would be created for both the educated and uneducated employed persons. Farm incomes will improve and gainful employment opportunities would expand."

On the other hand, the operations of MNCs have, of course, been beset with shortcomings and disadvantages. Their technical, financial and market network strength as well as risk bearing capacity can compete out the indigenous goods and services from the national market. For instance, under GATT provisions, improved and hybrid seeds can capture the Indian market at the cost of National and State level Seed Corporations and Private Domestic Seed Companies.

Similarly, these can capture (they have already captured) the business of processing of agro-products and considerably corner the domestic market on the strength of their quality products based on state of cost technology used (Pepsi taking over manufacturing of potato chips). To the extent the national companies are crowded out by the multinations, they develop the strength for treating the agricultural produce market as a captive market for the supply of raw materials.

These imminent dangers steering into our eyes should not diminish our efforts and also our courage to continue with the process of reforms. On the other hand, thoughtful strategy needs to be evolved to dilute the adverse effects and create workable channels of sharing the emerging global gains in a more sounder fashion.

The effect of MNCs entering into agro-based industries would be commercialisation of agriculture. A likely shift in the crop pattern from food crops to cash crops may be expected mainly in the areas served by irrigation. Another possible shift will be in favour of crops used for processing industries like maize and barley which are used for making breakfast cereals (Kallog's Corn Flakes).

As the food processing industry will aim at getting assured supply of inputs which will be agricultural products, the shift in crop pattern will be more in the areas served by better infrastructural support and assured irrigation facility. This may effect the food production adversely.

It may generally be expected that the large and medium farmers will be the first to adopt the changes, as they will be able to accept the risk involved. This may further increase the class differences in the rural society.

Commercialisation of agriculture as a result of new investment made by MNCs may lead further regional and crop imbalances which are highly visible in the post-green revolution scene of Indian agriculture. Crops which will go directly for processing and the areas where they will be grown will prosper much faster than other areas.

As a result of free entry of MNCs Indian agriculture may change to a commercial venture – Production as well as productivity of crops will have a chance of improvement. However, there is possibility that most of the investment may take place in areas served by better infrastructure and relatively well off part of the rural community will participate in the changed atmosphere, which may lead to further regional and income imbalance in the economy.

Employment in agriculture proper may decline over a period of time due to increase in mechanised farming, however, with the growth of the economy, employment situation is likely to improve. Many well known foreign companies have already entered and are showing interest in investing in India and many more may follow.

On the whole the policy changes will be beneficial for the development of agricultural sector and inturn the entire economy. Care will have to be taken that the new initiatives in agriculture bring about a rational balance between the growth of crops, on the one hand and among the regions as also among different systems of farming on the other. The exicitement of new regime should not blur the view of our own priorities, even when it would mean a relatively slow transformation without, of course, the processes of change which are immediately warranted to usher in a better state of economy on the threshold of next millennium.

6

Marketing in Agriculture

The marketing of farm products is a matter of great interest to the farmer, the consumer and the middleman. To the farmer, it provides a channel of communication between him and the society and gives him continuous information about the demand for his product. The consumer views it as a means of supplying his needs since marketing helps in raising the standard of living of people by satisfying a multitude of needs and desires of the consumers. The middleman depends upon it for his livelihood. These diverse interests lead the farmer to seek a high priced market for his products, the consumer a low priced market and the middleman, a margin between the farm price and consumer's price that will amply reward him for his service. The basis of agricultural marketing is that farm products are stored, transported to a place where wanted, at the desired time and transferred to those desiring them. It is for that reason that marketing has become part and parcel of the modern production process.

"Market reform", therefore, "ought to be an integral part of any policy for agricultural development. Normal economic incentives for farmers to increase their productivity can operate only to the extent that the marketing system enlarges the market for their produce and brings them a reasonable price for it. Their desire to earn large cash incomes can be stimulated by a marketing system that brings them cheap consumer goods, and their effort to increase productivity can succeed only if the system delivers the needed inputs. Marketing is as critical to better performance in agriculture as farming itself and should be treated with equal care."

Definition of Marketing

The objective of all economic activities is the satisfaction of human wants. In order to achieve this objective, the manufacturers, miners and farmers undertake production of good and services.

But they will fall to achieve their objective if the goods and services as produced do not reach the consumers who demand them at a time when they need them. It is marketing which helps making goods useful to the society by getting them where they are wanted, when they are wanted and by transferring them to those people who want them. Marketing in this sense, has been defined as "all the activities involved in the creation of place, time and possession utilities." To emphasise all these aspects of marketing, Clark and Clark wrote that, "marketing consists of those efforts which effect transfer in ownership of goods and care for their physical distribution."

According to Richard Kohls, "Marketing is the performance of all business activities involved in the flow of goods and services from the point of initial agricultural production until they are in the hands of the ultimate consumer."

A comprehensive definition of marketing has been provided by the Definitions Committee of the American Marketing Association. According to the Committee, "Marketing includes all activities having to do with effecting changes in the ownership and possession of goods and services. It is that part of economics which deals with the creation of time and place and possession-utilities and that phase of business activity through which human wants are satisfied by the exchange of goods and services for some valuable consideration".

In the light of the above definitions of the concept "marketing, we can now define agricultural marketing in a simple language. Agricultural marketing is a process by which the producers and buyers of agricultural goods are brought together. Such a definition emphasises two things (i) the buyers of agricultural goods, and (ii) the chain through which buyers and sellers of agricultural products are connected with each other.

Buyers of agricultural commodities can be divided into three main categories: (i) those who purchase agricultural goods for direct consumption; (ii) those who demand agricultural goods for use as intermediate goods; (iii) those who buy agricultural goods for sending them abroad, i.e., exports.

The chain which connects buyers and sellers of agricultural goods has so many links. For example, arrangements have to be made to collect agricultural goods from the sellers in the rural areas and store them before being sent to the market for sale. At mandies too, godowns have to be arranged till the stocks are disposed of. Two other important links of this chain are transport and credit facilities.

Significance of Agricultural Marketing

With the advance in intensive cropping, processing, storage and marketing of agricultural produce are becoming more and more important. In the absence of sound marketing facilities, the farmers have to depend on local traders and middlemen for the disposal of their farm produce which is sold at throwaway prices. In the case of small and marginal farmers, marketing of produce is still more critical. Since the amount of produce they are able to sell is small and overhead expenditure on processing, transport and marketing of the produce is heavy, the marketing of their produce is hardly remunerative if they do it on their own. Owens and Shaw very aptly pointed out, "Agricultural development is first of all a human problem, not a technical problem.

If all farmers have access to production inputs, the financial system, the market and agricultural knowledge, then they can improve the state of agriculture. But most farmers lack access to the market system and thus lack both the resources and the incentives to modernise their production methods." The authors add, "Unequal access to the market is one of the reasons why the Green Revolution in India has benefited mostly large farmers."

Thus, for the development of the agricultural sector, it is essential to develop agricultural marketing so as to match with the production surplus resulting from technological innovations and exploitation of the existing land and water resources. Apart from increasing production, marketing and distribution can significantly help in improving the availability of goods and services and also provide a stimulus to greater production. Had it not been for modern marketing, probably the Englishman would not have had his breakfast eggs from Denmark, cornflakes from Canada, and a cup of tea from the East. Marketing also increases demand which will provide its own incentive to increase supply.

This point can be illustrated by taking an example. Production of peas, for example, is seasonal and is limits to certain areas in the country because of the climate factors. But if the national market for green peas is developed, a guaranteed price and a stable income will provide an incentive to the farmers to grow peas on a larger area. Farmers will also take to planting of improved varieties of peas which will not only be better yielding but also will be frost resistant.

We, thus find that a scientifically developed market on a large scale for just one vegetable like peas could go a long way towards improving agricultural practices over a period of time. We may here quote Ashok Mehta who, while delivering his speech to the Indian Society of Agricultural Economics, said, "A revolution in agriculture is unthinkable in terms of something happening in agriculture alone.

It is to be thought of as a series of interchanges between agriculture and industry with rising intensity, industry supplying the basic needs of material inputs of agriculture, agriculture feeding back its surpluses for the development of industry, and industry supplying back the various consumer goods on which the agricultural surpluses can be spent. While emphasising this thought, I would like to go one step ahead and venture to say that the adoption of modern techniques of production has got to be integrated with the modern techniques of marketing and distribution, because in the ultimate analysis, the key to prosperity depends upon the availability of goods and services".

Special Problems in Marketing

Marketing of Agricultural Goods is more Complicated than the Marketing of other Non-agricultural Goods for the following Reasons:

1. The output of agriculture is largely a raw material which is used for further processing. This processing may be simple as in converting livestock into meat or it may be highly complex as in converting wheat into flour. Regardless of the complexities, it is true that in most cases, the product sold by the farmer soon loses its identity as a farm product and simply becomes "food".

2. Compared to most other products, agricultural products are both bulkier and more easily perishable. When things are bulky, marketing functions concerned with their physical handling are very much affected. In the case of those products which occupy a lot of space in relation to their value, the unit cost of transportation and storage is automatically raised.

   Similarly, many agricultural products are exceedingly perishable. The degree of perishability, no doubt, varies from product to product but perishability remains a distinct characteristic of agricultural goods.

These characteristics have their effect on the facilities necessary to market farm products. Bulkiness plus varying production, necessitates large storage capacities and higher cost of transportation. Perishable products require speedy handling as well as special refrigeration. Due to these characteristics, agricultural marketing sometimes is costlier than the marketing of industrial goods.

3. Much of agricultural production is highly seasonal. Most of the agricultural goods are ready for the market in certain fixed periods of the year only. For example, the harvest season of rice in India extends to two to three months but demand for it is evenly spread throughout the year. To the extent the commodity can be stored, the facilities must be provided to hold the product, but at peak season, the stores will be used at near capacity and at other times, they may be almost empty. If the product cannot be stored, it must either be processed or consumed immediately. Due to seasonal production, the processing plants will be running at full capacity during some periods and well below capacity in other periods. In case the product must move directly into consumption, transportation and refrigeration facilities must be immediately made available. All of these special situations mean increased costs in the agricultural marketing process.

4. The quantity as well as the quality of agricultural products are beyond the control of the producer. During some years, the growing conditions are such that the crop in general may be of high quality and more in quantity. In other years, unfavourable conditions may result in lower quality and lower production. Variations in quality and quantity of agricultural goods give rise to the following problems:

(a) Due to imbalance in supply and demand of agricultural goods prices fluctuate from time to time.

(b) Variations in the quality of products make it very hard to apply uniform standards for grades from year to year.

(c) Variations in quantity of agricultural goods throw transport and storage arrangements out of gear.

5. Agricultural output comes from many small units operating independently. Since the product is made available in relatively small amounts in the hands of many individual owners, the concentration process becomes very important in the case of agricultural marketing.
6. In the case of most agricultural goods, the demand is relatively inelastic, consequently, their prices rise steeply during the periods of short supply and fall sharply in the period of excess supply. It is for this reason that marketing of agricultural goods has to face the problem of price stability.

Characteristics of Agricultural Goods

The main characteristics of agricultural goods which make agricultural marketing much more elaborate and complex has been depicted in the following figure.

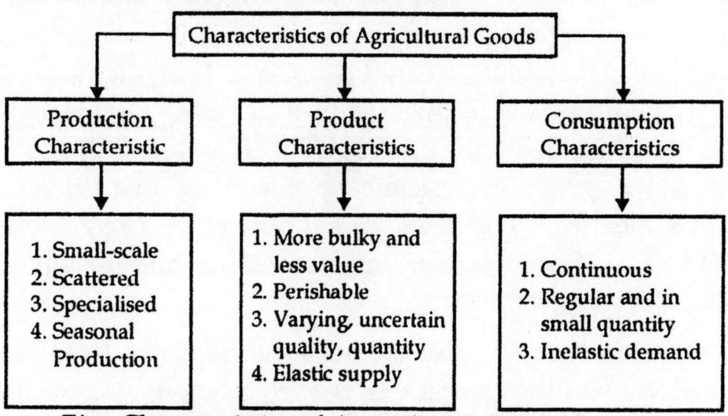

Fig.: Characteristics of Agricultural Goods Creating Problems of Agricultural Marketing

Marketing Functions

"Marketing function" may be defined as a major specialised activity performed in accomplishing the market process. Modern marketing system has several functions. Its important function is to move the desired varieties of farm products to consumers in the desired forms and conditions at the lowest possible cost. Different writers have used different methods of classification of marketing functions and services and somewhat different definitions of similar terms.

There are three principal marketing functions: (1) assembling (procurement, concentration), (2) processing (preparation for consumption), and (3) dispersion (distribution). In carrying out each of the three major functions of marketing certain other functions are performed which are termed as "secondary services". They include: (1) grading; (2) packing; (3) transporting; (4) storing; (5) financing; (6) assuming risk; and (7) selling.

We will now discuss briefly the principal as well as secondary market functions:

Assembling

In agriculture, goods are mostly produced by small farmers scattered over a wide area. Collection of small supplies from them, before further processes of marketing, is undertaken, is necessary. Assembling, therefore, means bringing together, collecting and concentrating goods of the same type from the various sources of supply at centrally located places. Agricultural goods are assembled chiefly for two purposes; first, for meeting the demand of consumers, and secondly, to provide a sufficient volume of business to middlemen like wholesalers and retailers.

The importance of assembling as a means of facilitating the orderly feeding of markets is fairly obvious where these markets are far away from the multitude of small producers responsible for supplying them. In a country like India, where about 70 per cent of farmers are small and marginal, this function has an important role to play.

Processing

We find very few agricultural products which are ready for final consumption when they leave the farm. In most cases, the marketing system must convert them into a suitable form before they can be sold to consumers. Further, excess farm production sometimes poses a serious problem of surplus over the existing demand in the coming years. Processing helps to create a new demand and maintain the quality of the product for a longer period. Processing, therefore, may be defined as the act or series of acts by which a product is converted into a more usable form.

The processing function would include all of those essentially manufacturing activities that change the basic form of the product, such as converting live animals into meat, fresh peas into canned or frozen peas or wheat into flour and finally into bread. To what extent processing can be regarded as a part of marketing is a debatable question. It is often not included in the list of marketing functions since it is essentially a form changing activity. But in order to provide consumers with the kind of food and clothing, when and where wanted during the year and to stabilise the market for farmers, processing is necessary. The function of processing which should have been carried out by the farmers themselves has been nowadays increasing delegated to non-farming specialists.

Dispersion

Dispersion is a process exactly the opposite to that of assembling. After collecting the products of many farmers in scattered localities and processing the same, the process of dispersion begins. This is dispersion of these products to many thousand of consuming markets and into the hands of million of consumers. Dispersion function involves finding where potential buyers are located, how much and what products they prefer, and what price they are ready to offer. It also includes selling of the goods, their physical movement and handling and the transfer of funds back to central and assembly markets. "It means keeping a steady flow into consumption of the vast volume of goods which is flowing into the central markets through the assembly end of the marketing system."

Secondary Services

These services are essential to the main functions of assembly, processing and dispersion. They facilitate operations all along the route of products through the market system. The main secondary functions are as follows:

Standardisation and Grading

Standardisation and grading imply the setting up of the basic standards to which the produced goods must conform. A standard specifies what basic qualities a product must have to be designated consistent with established characteristics. Standards are set with regard to the shape, size, colour, quality and performance, etc. The standardisation of products would, therefore, mean that the goods produced would ensure the same specifications as prescribed.

Standardisation, therefore, carries the idea of uniformity of quality. In its general sense, standardisation "includes the establishment of standards, the sorting and grading of products to conform to these standards."

The function of standardisation relates mainly to manufacturing products or can be applicable to agricultural processed goods. It is not possible for the farmer to conform to the same size, quality, colour, etc. In agricultural marketing, grading is very important. Grading is the act of separating or inspecting agricultural goods according to established specifications. It refers to the application of basic descriptive standards such as size, colour or appearance to agricultural products where the farmer has limited control over their products' physical specifications. For example, a fruit grower will command a lower price for a mixed lot of ungraded apples than the same lot after it has been graded into lots of equal size and appearance and priced accordingly. The purpose of grading is to establish a common language understood by buyers and sellers as a basis for judging the quality of a product in relation to its sale price.

Grading is also necessary to cater to the special tastes and preferences of different sections of buyers. Quality of a product and its utility are subjective. What standardisation can do is to offer uniform lots for sale in the market and to afford an opportunity to the buyers to purchase from the lot they like the best.

Packaging

Packaging is a process of putting the commodities into convenient containers or wrappers. Most of the products must be put into containers or packages to prevent spoilage and waste and to facilitate their movement from the farms into possession of final consumers. Packaging contributes to more efficient marketing by (1) reducing bulk; (2) facilitating handling; (3) reducing shrinkage and spoilage; (4) facilitating quality identification and product selection by consumers; (5) assisting in advertising and better merchandising; (6) helping to reduce other marketing costs by facilitating self-service retailing and modern handling methods through the marketing system.

In the face of widening agricultural markets, the packaging of agricultural products has assumed greater importance. Apples of Kashmir have a wide market and if not packed properly, run the risk of spoilage before reaching the final consumer. Some varieties of Indian rice enjoy a worldwide market and it should not be hard to understand the importance of packaging in this regard. Equally important are the changing perceptions of consumers and to satisfy them, packaging and constant improvements in it assume greater significance.

Transportation

Transportation is one of the most important functions of the modern marketing system. This function is primarily concerned with making goods available at the proper place resulting in the creation of place utility of agricultural commodities. In order to maximise the returns from an agricultural pursuit, it is not only the place utility but time utility as well that shall have to be kept in view. While it is always desirable to transport an agricultural commodity as far away as possible to a more remunerative market, it is equally important to reach these products to the consumers at the proper time. In both the cases, transport plays a crucial role. An efficient transport system enables to reach the markets far and wide and also without losing any precious time. Himachal Pradesh in India is an excellent example in this regard.

More than anything else, it gives transport top priority, realising fully well that an efficient transport system would facilitate the quick movement of agricultural and horticultural products from the farms to the consuming centres. In order, therefore, to ensure adequate returns to the producer, the transport system must be developed rapidly.

Storage

In respect of agricultural products, production is seasonal while consumption is throughout the year. Hence, storage of goods is imperative so that they can be protected from deterioration and their supply is ensured continuously. Storage, in fact, creates time utility. It implies storing of farm products in a warehouse or in a cold storage plant, depending upon the nature of the product.

Farm products are stored to make them available the year round, to build stocks during periods of plenty and release during periods of scarcity, as agriculture is characterised by relatively large and irregular seasonal and year-to-year fluctuations in production. Hence there can be two important reasons for holding agricultural products in storage. One is to even out the seasonal factor in production or in sales; the second, some agricultural products get improved in their quality as well as value when stored for a long time.

For example, products such as whisky must be conditioned to improve their flavour and thereby to increase their value. Bananas are plucked when raw and stored near the market site till these enter the final market.

This facilitates their transportation to distant places. Several other fruits and agricultural products are handled in the same way in order to minimise the risk of loss.

Financing

The financing function of marketing involves the use of capital to meet the financial requirements of the agencies engaged in various marketing activities. Farm products are not sold immediately after they are grown. It takes time for goods to reach the hands of the ultimate consumers. Thus, there is a long interval between the time when the farm products are ready for sale and the time when these goods reach the hands of the user.

To the extent that there is a delay between the time of the first sale of raw products and the sale of finished goods to the ultimate consumer, capital remains tied up in operation. "The service of providing the credit and money needed to meet the cost of selling merchandise into the hands of the final user is commonly referred to as the finance function in market."

Risk Bearing Function

The risk-bearing function is the accepting of the possibility of loss in the marketing of a product. These risks can be classified into two broad categories: (1) physical risks, and (2) market risks. Physical risks are those which occur from destruction of the product itself by fire, accident, earthquakes, cold and heat, etc. Risk attached to such natural hazards can often be transferred to institutions that specialise in assuming such risks. Insurance companies cover such risks in return for premium payments.

Market risks, on the other hand, are those which occur due to the changes taking place in product prices. Risks of changes in market demand are reduced through accurate sales forecasting and market research. Such risks can also be reduced through aggressive programme of advertisement, personal selling and the like.

Market research, in particular, is assuming considerable importance these days. Sales forecasts are made by specialised agencies and instructions are issued to the concerned parties accordingly. Market intelligence is also helpful in minimising the agricultural market risks. As the risk component is controlled, the marketing of agricultural products will become more efficient.

Selling

Selling or making persons to demand and agree to make payment for what a given seller has for sale is the central step around which all other marketing services revolve. On that score, it may be considered as the nucleus of marketing. Selling includes transfer of title and collecting or receiving of payment. Selling in its broadest sense not only effects ownership transfer but also helps in identifying prospective customers, stimulating demand and providing information and services to buyers. In order to achieve these goals, the market must combine such activities as personal selling, advertisement, sales promotion, packaging and custom services.

Marketing Margin

In dealing with the price problems of marketing, it is important to distinguish between the consumer price, the price or margin which marketing agencies receive for the services they perform and the price received by the producer of farm products. Each of the marketing services just described above, as well as the job of buying and selling, adds to the cost of the farm product by the time it reaches the consumer. The difference between the amount consumers pay for the final product and the amount producers receive is generally referred to as "marketing bill" or the "marketing margin". This margin between farm prices and retail prices reflects the cost of marketing and it varies widely for different groups of products.

The study of marketing margin of agricultural products is very important to ascertain the producer's share in the consumer's rupee and also to know the margins of various functionaries involved in the marketing process. It also helps us to know the different components of marketing margins and their influence on the final price which the consumer pays. Such a study for different commodities at different times helps us to know the variations in marketing margins.

This is essential for the formulation of an appropriate price policy for agricultural commodities. Marketing margin studies also assume importance because they help ensure reasonable returns to the producers and also make available the commodities to consumers at reasonable prices. Both from the point of view of the producer and consumer, such studies should help identity the areas where market charges can be minimised.

While it is very difficult to measure exactly the cost of marketing, many official and non-official studies made in many countries have confirmed the contention that distribution costs are relatively higher in agriculture. It is often said that the farmer's share in the consumer's rupee is relatively small as compared to that of several intermediaries in the market channel. The merchants absorb an alarmingly large share of this price. Even in England, where agriculture has been fully commercialised and the farmer does not lack business capacity, "the spread between producers' and consumers' prices", according to the Linlithgrow Committee, "is unjustifiably wide. Taken as a whole, distributive costs are a far heavier burden than society will permanently consent to bear." In the United

States, the farmer gets only half the price paid by the consumer for foodstuffs and in Germany about two-fifths.

In the USA, according to a study conducted by USDA, over the period of about fifty years (1915-1964), farmers have received an average of 42 per cent of the consumer's food dollar, while the marketing agencies have received 57 per cent of the food dollar.

Evidently, this is on account of inefficient marketing system which may arise out of two circumstances. Firstly, the high cost of marketing services and functions, system of transportation, wasteful methods of storage, careless handling of produce, large number of middlemen, etc. will certainly increase the cost of the products and deprive the producer of his due share in the consumer's rupee. Secondly, even if the marketing services are rendered most economically, the producer may be deprived of his due share on account of a defective market structure. Important reasons responsible for the high distribution cost of agricultural products may be listed as follows:

1. Difficulty of assembling the small outputs of thousands of producers scattered all over the country;
2. Greater incidence of transport costs, especially for bulky and low-priced products;
3. Difficulties of grading due to the wide variations in type and quality of output both of livestock and of crop products;
4. Impossibility of maintaining regular production and supply.
5. Greater need for storing and processing to cope with season quality of production;
6. Grater risk resulting from the need for storage and from the handling of highly perishable products; and
7. Strict limits within which sale by contract or by sample is possible.

Market Structure

In the case of agricultural industry, the farm products are produced by numerous small farmers and it may not be possible for each farmer to sell his produce to the potential consumer himself. The farmer producing on a small-scale has usually neither the time nor the ability to undertake directly the marketing of his produce. His produce is too small for him to undertake economically the direct marketing of his products to the final consumers. In most cases, the farmer's entire output is turned over to one or a small number of agent middlemen for marketing his produce. It is common for the bulk of commodities to change hands three or four times between the producer and the consumer.

Usually, farmers sell their produce to small merchants and travelling traders at the village level. This then is taken to the wholesale markets where it is again sold to retailers through which it reaches the final consumer. Farmers with bigger holdings may bypass the village merchant and sell directly in the major wholesale market. Certain commonly encountered and easily recognisable types of markets which generally exist between the farmer and the final consumer are:

1. Local assembly and processing markets;
2. District concentration and processing markets;
3. Primary, terminal or central markets;
4. Seaboard markets;
5. Secondary processing markets;
6. Wholesale distribution markets; and
7. Retail markets.

Local Assembling and Processing Markets

Before the supply of any agricultural product can *be* equated to the demand and distributed to the competing buyers, the available supplies of the product will naturally have to be collected from the innumerable farmers who produce it.

Most farm products pass through some small markets located close to the point of their production where produce is sold by the farmer to the local buyer who assembles the products of a number of farmers, and sends them to either the district or central market. These local markets, also known as growers' markets or primary markets, are located in small towns and at other convenient places where agricultural producers can bring their goods and sell to the prospective buyers.

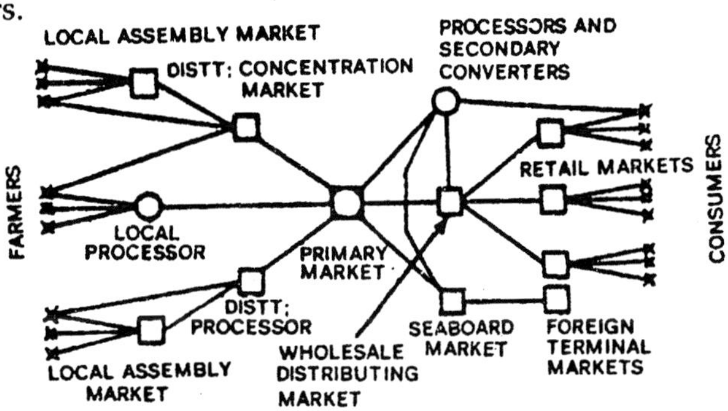

Fig: Types of Markets for Farm Products

District Concentration and Processing Markets

Some commodities are processed either wholly or in part at local markets where consumers purchase them directly from the farmers. But most of the commodities are processed at the large central markets. Between these two types of processing markets, in case of some products, we find district processors or markets where the produce from a considerable number of local assembly markets are assembled and processed for transporting to central markets or direct to consuming markets.

Central Markets

In the previous figure we find that in between the fan shaped producing and consuming ends of the market system, there are large central markets where goods from many local assembly markets or district processing points are concentrated for processing or for additional processing, storage, grading, conditioning and distribution to other processing and secondary wholesale markets or to markets abroad. Central markets, thus, are the last step in the assembling or concentration and the first step in the dispersion process. The big wholesale markets can be compared to reservoirs into which flow the fluctuating and seasonal supplies from different producing areas, and from which is met a constantly fluctuating demand. It is in this market where price-making activities take place.

It is in these markets where buyers and sellers representing regional, national or even international demand for the agricultural products are seen more clearly in the wholesale market than in producing areas. The adjustment of demand and supply may Le said to take place in the central markets. As a matter of fact, a central or terminal market is the meeting point of the three marketing processes – concentration, equalisation and dispersion.

Seaboard Markets

Seaboard market specifically serves the place for ocean shipment to foreign centres. These markets either receive the agricultural goods from the central market or directly from the locai and district assembling markets. These markets have facilities for unloading storing, loading ships and rendering incidental services such as grading and price reporting. Corresponding facilities for handling imports of agricultural goods are also available in such markets.

Wholesale Distribution Markets

The agricultural goods which are now assembled in the central market have to be despatched to consumers through the retailers. Wholesalers operating in the central markets mostly prefer to deal in bulk which cannot be afforded by retailers. Therefore, separate wholesale markets specially for the dispersion of goods to retailers have been developed in between the large central markets and the retailers. These markets are often called "secondary wholesale markets" to distinguish them from primary or central wholesale markets. Another term sometimes applied to them is "jobbing markets".

These markets operate on a smaller scale than the central markets. They do not play any major role in the determination of prices. They are concerned only with those operations which are essential to meet the needs of their retail clients.

Retail Markets

In retail markets, the goods are ultimately placed before the individual consumer for acceptance or rejection. Every retail store in the city, town and village which sells to the final consumers is a retail market. "Retailing is, perhaps, the most difficult part of the marketing process to perform, and certainly is the most expensive. For producers of consumers' goods, the retail store is the ultimate point of contact with the users of their products. Here in the retail store, the battle of the market is fought out to a final conclusion."

Cooperative Marketing

Various marketing functions, such as collection, storage, financing, insurance, grading, transporting and selling of agricultural produce are performed by a large number of persons who play the part of middlemen. Very often, there are more middlemen than are necessary and generally the charges are out of proportion to the services they render. The result is that the primary producer does not receive his due share of the price paid by the ultimate consumer. A bulk of price paid by the consumer is digested by these middlemen with the result that the consumer has to pay a higher price while the producer gets a lower return for the same. Cooperative marketing of agricultural produce assumes significance from the need to reduce the price spread between the producer and the consumer so as to ensure a better return to the primary producer without affecting the interests of the consumers. Cooperative marketing is the organised sale of farm products on a non-profit basis in the interests of the individual producer. Such cooperatives might start either from the producers' end where they may attempt to distribute their own products in the hope of obtaining higher returns, or from the consumers' end where they may take over distribution in order to buy cheaper.

It is well to stress that the real purpose of cooperative marketing is not to eliminate the middleman but to perform his services at a lower cost.

The primary objective of cooperative marketing is to reduce the marketing margin and thereby assure the farmer a better price for his produce and supply the goods to the consumer at a fair price. A cooperative marketing society may have the following objectives".

1. To make arrangements for the sale of agricultural produce of members. (It may deal with non-members also but they will not get those benefits which are available to members.)
2. To provide loan facilities to members on the security of agricultural produce.
3. To give encouragement to members to grow the latest improved and standardised varieties of produce.
4. To make arrangements of godown for the preservation of produce of members.
5. To make arrangements for processing and cold storage activities for members.
6. To encourage thrift, self-help and cooperation amongst members.
7. To undertake those other activities which can further the objects of the marketing societies.

Advantages of Cooperative Marketing

When the farmers combine and set up a cooperative sale society to collectively market their produce, they are able to enjoy a number of advantages. These are:

Reduced Cost and Improved Service: In the absence of cooperative marketing, assembling is done by a number of local dealers each of whom does not have a sufficient volume of business to reduce his expenses of marketing to the minimum. A cooperative society which includes a large number of farmers in a given area might be able to control a large volume of business and thus effect real economies in services such as transport and storage. In this way, a cooperative marketing society can provide the farm producers improved services at reduced cost.

Improved Marketability of Products: A cooperative society is in a better position to study the needs of the consumer than the individual farmer. Its main achievement would be to make products conform to the needs of the market. Further by providing better grading facilities, a cooperative marketing society can induce its members to produce those varieties of the product for which there is a greater demand in the market. Thus by paying its members according to grade, it induces the farmers to "send the right things to the market at the right time and in right Quantities."

Improved Bargaining Power: Even if a cooperative marketing organisation only takes over the functions of country dealers, it improves the bargaining position of farmers and makes it possible for them to negotiate, on an equal or perhaps even a superior footing, with large-scale wholesalers. In the case of cooperative marketing, the buyers of agricultural produce have to deal with a farmers' association which controls a large part of the limited supply. Further, the society as a representative of the farmers, can present the farmers' case much better to outside agencies and secure for the farmers the necessary concessions.

Steadier Supply to the Wholesale Market: The cooperative society may be able to offer to sell the wholesalers a more regular and assured supply of the produce than they can obtain if they have to compete with other wholesalers for the produce of many farmers.

Controlling the Flow to the Market: The formation of a cooperative society makes it easier to adjust producers' supply to demand. It can take over the produce from the growers when the harvest is over, but release it gradually in the market according to demand. In addition, the society can inform the farmers as to the probable future trend of prices, and advise them whether to increase or decrease their scale of production. It is much easier for the cooperative societies than for private firms to educate their members as to what consumers want and when they want it.

Development of Markets: Cooperative marketing societies are in a better position to develop the market for their produce by effective advertisement and other means. For an individual farmer, advertisement will obviously be beyond his reach.

Cheaper Finance: These societies can provide loan facilities at a cheap interest rate to members on the security of this agricultural produce. Cooperative society itself can secure a loan on the security of produce stored at much lower rates of interest than the individual farmer.

Training in Business Methods: Finally, cooperative marketing has the merit of giving the farmer an insight into the marketing mechanism. He starts understanding the problems and difficulties of the middlemen and sees that they may not be exploiting him. "Experience in a marketing association gives farmers a more rational outlook on marketing methods and finally, it gives them business experience and a higher sense of business responsibility."

All the above advantages of cooperative marketing can be secured only if the following conditions are fulfilled:

1. In order that the society can reduce its cost of marketing, bargain effectively in the wholesale markets, raise necessary marketing finance at cheap rates and effectively control the flow of the commodity to the markets, it is necessary that it secures a large volume of business.
2. Secondly, the cooperative marketing society should be able to secure efficient management. "The business of

sale is not a matter for amateurs .... skilled technical advice and guidance are necessary but even more important is the proper business management. Unless this is available, the society must inevitably come to grief'.

3. Thirdly, in order to carry on its business effectively, the cooperative must have sufficient capital at its disposal. "In a country with a well-developed banking system, where marketing finance is available at low rates, where there is a well-developed system of advancing loans on the security of warehouse receipts, a cooperative marketing society has greater chances of success."

Cooperative Marketing Societies

The main job in agricultural marketing is how to take over the responsibility of the produce of the farmer and give him the best of benefits and amenities without subjecting him to the intricacies of market transactions. In this respect, the cooperative marketing societies were being established prior to 1954 but since then multipurpose societies have been started which, on the one hand, provide credit facilities to the farmers and, on the other, undertake the responsibility of marketing their surplus produce.

As soon as the members of these societies supply surplus agricultural produce to these societies, they get an advance to carry on their agricultural operations. The agricultural produce thus collected from members and non-members is processed and then brought to the Mandi for disposal. If current prices are not favourable, society may decide to stock the produce. As soon as the produce is sold, farmers are paid their remaining share after deducting the advance already made.

A significant experiment is being conducted by organising Farmers' Service Societies (FSS) in pursuance of the recommendations made by the National Commission on Agriculture. These societies would disburse all types of credit required, supply inputs including fertilizers, arrange for processing and marketing wherever possible, and undertake all connected activities directly or in conjunction with other organisations, Necessary steps are being taken by the state governments to organise at least one such society in each of the districts covered by special programmes like SFDA and MFAL Development Agencies, Drought Prone Area Programme and the Command Area Development Programme.

Progress of Cooperative Marketing

The process of developing cooperative agricultural marketing in India was initiated during Second Five Year Plan on the recommendations of All India Rural Credit Survey Report. The Committee's main recommendations were State participation in the share capital of marketing societies at all levels, linking credit with marketing, establishment of chain of warehouses and management of cooperatives by well-trained personnel. A three-tier system of cooperative marketing is prevalent in India.

1. Primary cooperative marketing at mandi level.
2. District cooperative marketing society at district level, and
3. Apex cooperative marketing society at the state level. In some states, viz. Assam, Bihar, Kerala, Madhya Pradesh, Orissa, Rajasthan and West Bengal, only two tier system is working since in these states district level cooperative societies are non-existent. At present, the cooperative marketing structure comprises 2,633 general purpose primary cooperative marketing societies at the mandi level, covering all the important mandies in the country, 3,290 specialised primary marketing societies for oilseeds, etc. 172 district/Central Federations and the National Agricultural Cooperative Marketing Federation of India at the national level. The total value of agricultural produce handled by marketing cooperatives amounted to about Rs. 9,504 crore in 1994-95, as against Rs. 1,950 crores in 1980-81.

To conclude, cooperative marketing in agricultural products is of immense advantage to the farmers. It releases them from the strains of a parasitic middle agency that chops off a substantial share of profits. However, the success of cooperative marketing societies is greatly dependent upon the efficiency with which these operate. Usually, it has been found that such societies do not do well because of shared responsibility. On the organisational front, therefore, the management of cooperative markets must be sound, honest and efficient. By ensuring fair rewards to farmers through these markets, agricultural pursuits as a whole can be made more remunerative which will go a long way in boosting agricultural production.

Agricultural Marketing in India

Agricultural marketing system prevalent in India is characterised by a considerable degree of diversity and flexibility in the arrangements as between different commodities and regions. The marketing arrangements comprise, on the one hand, systems operated freely by private enterprise without any state intervention and, on the other hand, are systems which are completely under state control. In between these two extremes, other arrangements with varying degrees of government intervention and support in the areas of price fixation, procurement quota, buffer stocks, credit controls, etc. are also prevalent. The need for trying out so many varying marketing system arises from the greatly dissimilar conditions of production and marketing as between commodities and states. These have been evolved over a period of years on the basis of experience. These systems are also subject to changes and modifications as may be warranted from time to time by the changing production and marketing situations as well as economic policies.

Although state governments enjoy full autonomy in the matter of agricultural policies, including those relating to marketing, yet a broad framework of agricultural policies including marketing are laid down by the Central Government.

At the Centre, agricultural marketing is the concern of the Ministry of Agriculture and Rural Development.

At the state level, the administrative setup for agricultural marketing differs from state to state. In some states, agricultural marketing is in the hands of completely independent departments. In many others, the work is entrusted to a small section forming a part of the Department of Agriculture without a separate identity of its own. At the level of the state secretariat, the subject "Agricultural Marketing" wherever not handled by an independent department, is dealt within the Department of Agriculture or Cooperation.

A recent development is the formation of Marketing Boards in a number of states. The functions of these Boards vary considerably from state to state. While in some states, these Boards have merely advisory functions, in others, the Marketing Boards are gradually taking over the responsibility for all agricultural marketing activities in the state.

The system of agricultural Marketing as it stands now is complex and not neutral to returns to individual farmers. Though the state has come in a big way in this area, a lot more remains to be done to streamline agricultural marketing in the country.

Existing System of Agricultural Marketing

The existing system of agricultural marketing in India is as under:

Sale of Agricultural Product to Moneylenders or Village Traders: With the advances made in intensive cropping, processing, storage and marketing of agricultural produce are becoming more and more important. In the absence of sound marketing facilities, the farmers have to depend upon local traders and middlemen for the disposal of their farm produce which is sold at throwaway prices.

In the case of small farmers, marketing of produce is still more critical. Since the quantity of their produce is small and overhead expenditure on processing, transport and marketing of the produce is heavy, the marketing of their produce is hardly remunerative if they do it on their own. In most cases these farmers are forced, under socio-economic conditions to carry on distress sale of their products. In most small villages, the farmers sell their produce to the moneylender from whom they have usually taken loans, who in turn, passes it over to a bigger merchant or firm in the neighbouring market-town. According to an estimate, 85 per cent of wheat and 75 per cent of oilseeds in Uttar Pradesh, 90 per cent of jute in West Bengal, 70 per cent of oilseeds and 35 per cent of cotton in Punjab are sold by the farmers themselves in the village.

Hats and Shandies: Another prevailing method of disposing the produce is selling it in weekly or biweekly bazars called hats held in the village or a nearby town. Sometimes, produce is disposed of in the fairs which are held at some places on important religious occasions. The area covered by a hat usually varies from 6 to 12 kilometres. The sellers in these hats are usually marginal and small farmers. The big farmers with large surplus go to larger wholesale markets.

Primary Markets: The third way of disposing of the farm produce is to take it to primary markets. There are more than 22,000 rural primary markets in the country where petty dealers and agents of merchants operate. A great majority of these primary markets is owned and controlled by local bodies, a smaller number is controlled by regulated markets and a few owned by private individuals.

Wholesale Markets: From the rural primary markets, the activity shifts to the wholesale assembling markets of which there are more than 5,000 in the country. A majority of wholesale assembling markets is controlled by Regulated Market Committees at the local level and by the State Marketing Boards or State Marketing Departments at the state level. At these markets, a host of functionaries such as wholesale merchants, commission agents, brokers, etc. operate and aid the marketing process.

Defects in Agricultural Marketing

It is quite obvious that the present system of agricultural marketing in India is far from satisfactory. The following are the main defects:

Forced Sales: One of the major defects in agricultural marketing in India is the inability of the majority of the farmers to wait for long after harvesting their produce. The average Indian farmer is so poor and indebted that he cannot afford to wait till such time when he will be able to sell the produce at reasonable higher prices. In order to meet his commitments and pay his debt, he is forced to sell the produce just after the harvest at whatever price is offered to him. The Rural Credit Survey Report rightly remarked that producers in general, sell their produce at an unfavourable place and at an unfavourable time and usually they get very unfavourable terms.

Inadequate and Expensive Facilities of Transport and Communication: One of the important components of the infrastructure necessary for the growth of agricultural marketing is communication and transport. Forced and distress sales by the farmers are further facilitated by the fact that the means of transport available to the farmers are inadequate and very expensive. Even at present, there are many rural areas which have not yet been well connected with main roads or with marketing centres. Most roads in the rural areas are Kutcha (bullock-cart roads) and in the rainy season, no means of transport can ply over these. Besides inadequate transport facilities, they are so costly that they are beyond the reach of many small and marginal farmers. Difficulties also arise at the terminals, where exploitation of innocent and illiterate farmers often takes place.

Inadequate Storage Facilities: Storage facilities in the rural areas and in the primary markets are either totally absent or grossly inadequate. Farmers mostly store their surplus grains in earthen vessels or in the Kutcha rooms or in the underground Kutcha pits. This type of storing exposes the grains to white ants, rats, dampness, etc. and results in considerable loss. Lack of storage facilities also forces the farmer to sell immediately after the harvest.

With the increase in agricultural output, there is an obvious need for increased storage to minimise wastage. While production has maintained an upward trend, facilities for storage have not kept pace. The task now is to create additional facilities for storage of all agricultural products, more particularly foodgrains in the rural areas. In the absence of adequate storage facilities in the rural areas, the farmers are compelled to sell their produce immediately after harvest at prevailing prices which are bound to be low. Such distress sale of farm products deprives the farming community of its legitimate incomes.

Further, farmers are compelled to store foodgrains and other products in improvised substandard godowns. Such storage of foodgrains and other products leads to deterioration of these commodities owing to dampness, etc. As a result of this, the quality of agricultural produce goes down and the farmers are again compelled to sell their produce at a lower price in view of deterioration in quality. Storage in Kutcha godowns also leads to a loss due to the attack of rodents, insects, pests, etc. The Parse Committee estimated the post-harvest losses at 9.3 per cent of which nearly 6.6 per cent occurs due to poor storage conditions alone.

There is, therefore, great need for providing the farming community with storage facilities near their farms. These godowns should provide facilities for scientific storage so that wastage and product deterioration are avoided. The farmer should also be able to get his credit requirements without being compelled to sell his produce at a time when prices are low.

Superfluous Middlemen and Malpractices: In the absence of an organised marketing structure, private traders and middlemen predominate the marketing and trading of agricultural produce. In between the farmer and the ultimate consumer, there is a large army of middlemen who function at various stages in the process of assembling and distribution of farm produce. In the village, there is the bania; in the town, there are commission agents, brokers and others who get fat at the expense of the poor farmer.

Mukherjee, in this connection, says "the system of agricultural marketing in India is saddled with a long chain of middlemen and the remuneration for their services, increase the load on the consumer, although the producer does not derive similar benefit". Many market surveys have revealed that middlemen take away about 48 per cent of the price of rice, 52 per cent of the price of groundnuts and 60 per cent of the price of potatoes offered by consumers.

Unauthorised Deductions and Multiplicity of Charges: The sale proceeds of the farmer are subjected to many unauthorised deductions like Zakat, Dharmada, Goshala, Shagirdi, etc. In addition to these charges, a number of other charges, legitimate or illegitimate, are also claimed by the middlemen. Such allowances are claimed on account of moisture, dust, husk, etc. The Report on the Marketing of Wheat in India observes that, "not only the arhatia and dalal, but the minimum, the chowkidar, the sweeper, etc. regard themselves as entitled to share of his produce." Anyone connected with the marketing of agricultural produce directly or indirectly, nearly or remotely, legally or illegally, claims his share in the rupee that becomes due to farmers with a result that the poor fellow is back home with only a fraction of the rupee that his produce fetched him in the market.

Absence of Grading and Standardisation: Heterogeneity in quality of agricultural produce is a universal phenomenon. Majority of farmers in India sell their farm produce without proper grading. In such a "dara" sales wherein heaps of all qualities of grains (good and bad) are sold in one common lot, farmers are deprived of a better price for a better produce. In fact, the whole lot gets rated at the inferior level and yields the farmer a return much lower than the one he deserved.

Lack of Marketing Information: Most Indian farmers do not get the necessary information regarding prices of different farm goods prevailing in different markets. Farmers' most important source of information is the village bania or the middlemen whose information is more biased in favour of the market trader and hence our farmer fails to realise reasonable returns for his produce. The situation in this regard should have improved recently due to the expansion of the television network in different parts of the country. But due to lack of associated amenities, the farmer fails to get a full advantage of this facility.

Government Measures for Agricultural Marketing

A number of measures have been initiated by the Government of India to improve agricultural marketing since independence. The suggested lines of improvement along which the action has been initiated by the Government can be listed as follows:

Establishment of Regulated Markets: As mentioned above, a major defect in agricultural marketing in India has been the prevalence of unfair means and malpractices in most of the markets. These practices rob the fanner of his due share and place him at a disadvantage *vis-avis* the more organised section of the trading class. Regulation of agricultural markets aims at eliminating the unfair market practices not only of existing markets but the regulation of new markets set up under law. The regulation of markets generally introduces a system of competitive buying, helps in eradicating trading malpractices, ensures the use of standardised weights and measures and evolves a suitable machinery for the settlement of disputes, there by ensuring that the producers are not only not subjected to exploitation but also receive remunerative prices while disposing of their produce in the markets.

The enactment of Agricultural Produce Markets Acts has enabled the states to establish regulated markets administered by Market Committees on which are represented producers as well as traders. As on 31-3-1984, there were 2,045 principal markets and 3,534 sub-markets which had been brought under regulation. These markets have to grow in size and number to universalise the regulated markets in the country. With this, the farmer will get commensurate returns and a stimulus to grow more on his fields.

Provision of Storage and Warehousing Facilities: Scientific storage is crucial from the point of view of farmers and the consumers. At present, there are a number of agencies engaged in warehousing and storage activities. The Food Corporation of India, the Central Warehousing Corporation and State Warehousing Corporations are among the principal agencies engaged in this task. Besides, the ARDC is also financing suitable agencies engaged in building up storage capacity.

Table: Storage Availability with the Public Agencies
(In lakh tonnes)

Agency	*1973-74*	*1979-80*
F.C.I.	51	76
C.W.C.	12	20
S.W.Cs	6	24
State Governments	18	19
Cooperatives	32	47
Total	119	186

In addition to the above capacity, the FCI has taken over a hired capacity of 3.8 million tonnes from private parties.

Taking note of the inadequacy of storage facilities in rural areas, and the consequent hardships and losses to the rural community, the Central Government is implementing the scheme for the establishment of a National Grid of Rural Godowns since 1979-80. This scheme is different from the other schemes in as much as it is primarily meant for providing warehousing facilities to the farmers. The aim is to create an additional storage capacity of 2 million tonnes at the village level by the end of 1985. The capacity of the godowns vary from 200 to 1,000 million tonnes.

Only limited studies about the actual shortage of storage at the village level have been conducted so far. In a study conducted by the Administrative Staff College, Hyderabad in 1975, it had been worked out that the gross additional rural storage required as a percentage of production in Punjab will be 25 per cent and in Andhra Pradesh as high as 42 per cent. This is an indicator of the kind of additional storage requirement in the rural sector. In spite of the absence of precise figures of storage requirement in rural areas, a programme to meet the requirements of safety of foodgrains for orderly transportation and marketing and for ensuring farmers the return they deserve under the proclaimed policy of the government is called for.

The Working Group on additional storage facilities in rural areas has recommended a scheme of establishing a network of Rural Storage Centres to serve the economic interests of the farming community by providing them warehouses to store their agricultural products. The proposed scheme is intended to:

1. Prevent distress sale of foodgrains and other agricultural produce immediately after harvest at low prices and to strengthen the farmer's staying power;
2. Ensure remunerative prices to farmers by storing agricultural produce in well-managed godowns till favourable prices could be obtained;
3. Avoid loss and deterioration in quality while in, storage;
4. Provide easy credit facilities against the pledge of stocks;
5. Reduce pressure on storage space with existing public agencies;

6. Reduce pressure on transport facilities in post-harvest periods of peak demand for transport; and
7. Create additional opportunities of employment.

Grading and Standardisation: With a view to making available unadulterated and standard quality agricultural produce including horticulture and livestock produce, to the consumer and ensuring better prices to the producers, grading and standardisation of agricultural produce is carried out under the provisions of Agricultural Produce (Grading and Marketing) Act, 1937. Under this Act, grade specifications in respect of 142 commodities have so far been notified.

Three Categories

1. Compulsory grading for export;
2. Grading at producers' level; and
3. Voluntary grading for internal consumption.

In order to promote export trade of agricultural produce by ensuring quality products to the exporters, compulsory grading of agricultural produce has been introduced in respect of 41 commodities under "Agmark". The important commodities covered under this scheme are spices, unmanufactured tobacco, walnuts, vegetable oils, essential oils, onion, ginger, potatoes, pulses, etc. Under this scheme, the value of commodities graded for export during the year 1983-84 was Rs: 593.92 crores which is estimated to be Rs. 653.48 crores during the year 1984-85.

Voluntary grading of agricultural commodities under the Agmark for internal trade and consumption is aimed at ensuring products of purity and quality to consumers. There are, at present, 15,467 holders of certificate of authorisation under "Agmark" for grading of agricultural commodities in the country.

For securing an adequate return to the producers the programme of establishment of grading units in the regulated markets and marketing cooperatives has been introduced since

1962-63. For this purpose, a number of grading centres have been established in various states by the State Marketing Departments or Marketing Boards.

The total number of grading centres set up at producer's level in March 1984 was 808. Paddy, wheat, maize, jowar, potato, gram, tur, groundnut, mustard, mustard-seeds, dry chillies, fruits, onion, cotton and jaggery are among the important agricultural commodities graded at the producers' level. The value of commodities graded at the producers' level during 1984-85 was about Rs. 600 crores.

Use of Standard Weights: Using of non-standard and arbitrary weights by "arhatiyas" and brokers to cheat the farmer is one of the major defects of the unregulated markets. A number of steps have been taken by the government in this direction. The Standard Weights Act was passed as back as in 1939. After the introduction of metric system of measures in 1958, all old system of weights and measures have been replaced. Weights and Measures Departments in the states take all precautions to see that only standard weights are used in the regulated markets.

Dissemination of Market Information: In order to apprise the farmers of prices prevailing in the different markets and the condition of demand for different agricultural commodities, market information relating to agricultural products is made available to the farmers through different media like television, newspapers, radio, etc. Market intelligence reports are made available to the farmers and are displayed in a number of markets in the country. Information about the prevailing prices of important commodities in different markets is provided daily by All India Radio and in a regular manner on the television network.

Better Transport Arrangements: Another important infrastructure badly needed for agricultural marketing is rural transportation. Vegetables, foodgrains and products of animal husbandry which are being harvested in plenty nowadays cannot find their way to the nearby towns unless villages are connected by proper roads. Our planners have assigned higher priority to the construction of rural roads. According to the information available with the Planning Commission, 26,724 out of 69,400 (67.3%) villages having a population of over 1,500,26,814 out of 56,608 (47.4%) villages having a population of between 1,000-1,500 and 1,02,738 out of 4,67,075 (22.1%) villages having less than 1,000 population are expected to have been connected by all-weather roads as on 31st March, 1984. Efforts to develop more rural roads and to link many villages with the markets will be pushed up during the Seventh plan period.

Provision of Central Assistance: A scheme for providing Central assistance for the development of selected regulated markets is in force since 1972-73. Since the inception of the scheme and till 31-3-1984, 449 selected regulated markets have been granted Central assistance of Rs. 1,557.18 lakhs. To help the farmers, particularly the small and marginal ones, who depend on the nearest rural markets like "hats", "shandies", etc. for the sale of their small marketable surplus, a scheme for the development of primary rural and wholesale markets in backward areas is in force since 1977-78. Under this scheme, grant-in-aid to the extent of Rs. 1.50 lakhs per primary rural market and Rs. 5 lakhs per rural wholesale market situated in backward areas has been made. During the years 1977-78 to 1983-84, Central assistance of Rs. 2,224.30 lakhs has been granted for development of 1,948 rural primary markets.

7

Agricultural-Economic Transformation

Schultz's is the only significant contribution on the subject of the transformation of traditional agriculture. Here, we present a brief review of his outstanding piece of work.

The Problem

If production is continued on old and established methods, not much can be produced even if lands were rich and people worked hard. Application of science and technology, on the other hand, enables greater production even when the soils are poor and men devote less labour.

Farming that is carried on with the factors of production used for generations may be called traditional agriculture. Those countries which are dependent upon traditional agriculture are poor and spend much of their income on food. But when a country develops its agricultural sector, food becomes more abundant, incomes of the people rise and less of the income of the country is spent on food. "HOWL to transform traditional agriculture which is niggardly, into a highly productive sector of the economy is the central problem" of the thesis put forth by Professor Schultz.

The transformation problem, basically, is an investment problem. Investing in agriculture will improve the performance of this sector. Supply of investment may not be a serious problem but to determine the forms of this investment in agriculture is the real problem. The problem relates to making agriculture a source of economic growth which, according to Prof. T.W. Schultz, has not been done even though the literature on economic growth has voluminously increased. Many countries have sought to industrialise without giving comparable attention to agriculture. It is believed that modernised agriculture can make substantial contribution to economic growth of a country.

Delayed attention, to the economic potentials of agriculture has been due to economists' neglect of the agricultural sector in less development economies. Earlier development studies mostly took the form of macro-development models which were not relevant in theorising about the growth potential of the agricultural sector or examining the behaviour of agriculture as a source of growth. Labour and capital have been chiefly used as explanatory variables in agricultural production function, leaving the residual to technological change. "When these models are used to analyse real data, it turns out that most economic growth is concealed under the heading 'technological change'. Because it is a residual, there is no explanation for it. At the same time, those who have taken any notice whatsoever of the agricultural sector always about the backwardness of farm people." It has been widely held that traditional agriculture can improve only when farm people learn the economic virtue of "work and thrift".

Old Doctrines

Old doctrines did recognise the importance of agriculture in the process of economic development. Physiocrats believed that agriculture alone is productive, for it yields, the subsistence of its workers, the earnings of its entrepreneurs and a surplus (rent), while industry and trade are sterile since no such surplus (extra income) is produced in these two sectors.

The classical economists put forth a dynamic model of growth based on accumulation of capital, Malthusian population law and law of diminishing returns in agriculture. According to Adam Smith, "Food is the conditional factor in the growth of an economy. When by improvement and cultivation of land, the labour of one family can provide food for two families, the labour of half the society becomes sufficient to provide food for the whole. The other half, therefore, can be employed in providing other things, or in satisfying the other wants and fancies of mankind."

Schultz refutes all these doctrines. He believes that these doctrines do not stand empirical tests. Schultz also refutes the doctrine of "disguised unemployment" and factor indivisibilities.

Unsettled Questions

In determining the opportunities in agriculture for economic growth, according to Schultz, it will be necessary to resolve three unsettled questions:

1. Can low income communities increase agricultural production substantially by an efficient allocation of the agricultural factors of production presently at their disposal?
2. Which agricultural factors of production are primarily responsible for the large differences among countries in the success of the agricultural sector in contributing to economic growth?
3. Under what conditions does it pay to invest in agriculture?

The first and third questions are of fundamental importance and form the core of the problem of transformation of traditional agriculture. So far as the first question is concerned, the general belief held is that agriculture in less developed countries is inefficient in using the factors at the disposal of a farm. Professor Schultz contradicts this hypothesis. His study supports the proposition "that differences in land are least important, differences in the quality of material capital are of great importance, and differences in the capabilities of farm people are most important in explaining the differences in the amount and rate of increase of farm production."

In several countries such as Mexico and Japan, agriculture has done several times better than that of India, not because they have superior quality lands but they have improved skills and schooling of farm people and used better quality capital.

Attributes of Traditional Agriculture

Traditional Agriculture is Supposed to Possess Three Attributes:

1. Farming is a way of life based on long-established traditions. Traditional agriculture is, therefore, cultural characterisation of the way people live.
2. Another attributes of traditional agriculture is the institutional setup dealing with the ownership of land, legality of tenureship and the share of home consumption in agricultural production.
3. Another aspect is of technical property.

Not much importance has been attached to these aspects by Schultz. In his view, traditional agriculture is a particular type of economic equilibrium, it is an equilibrium at which agriculture gradually arrives over a long period, provided particular conditions prevail. This type of equilibrium is not readily subject to change.

Traditional agriculture, according to Schultz, can be transformed into a relatively cheap source of economic growth. He tries to show what such transformation entails and what means are required to accomplish it efficiently.

According to Schultz, new factors of production are needed to transform the traditional agriculture. Introduction of a new factor brings in some problems. Whereas past practices have to be discarded, on the one hand, new strategies have to be developed, on the other to meet the new situation arising out of risks and uncertainties involved in agricultural transformation. It is, therefore, not only significant to introduce new factors of production and reap richer harvests, but learning from experience, what risks and uncertainties are involved is also there.

The hypothesis to be investigated is, therefore, "that the rate at which farmers who have settled into traditional agriculture except a new factor of production, depends upon its profit, with due allowance for risk and uncertainty and in this regard the response is similar to that observed by farmers in modern agriculture."

Reallocating or readjusting the allocation of resources, according to Schultz, is not going to make any substantial improvement in farm income. Even if the imperfections of capital market are corrected, it is not going to open any sound opportunities of economic growth. Even when it is known that there is a superior technology available, the farmers may not be forthcoming to adopt it. This is a riddle.

Allocative Efficiency

The hypothesis Schultz set before himself is: "There are comparatively few significant inefficiencies in the allocation of the factors of production in traditional agriculture."

The factors refer to traditional factors of production in possession of a community. All poor agricultural communities may not be traditional and may have to be excluded because these have witnessed change. The implications of the proposed hypothesis are:

1. "That no appreciable increase in agricultural production is to be had by reallocating the factors at the disposal of the farmers who are bound by traditional agriculture". By further implications, it means that all factors of production are allocated with full regard to marginal returns and marginal costs.
2. Significant individualities will not hamper agricultural production.

3. No farm management expert shall be in a position to discover any major inefficiency in the allocation of factors.
4. No production factor remains unemployed.

According to Schultz, treatment of the proposed hypothesis, if any one of the implications mentioned above is proved wrong, the hypothesis of allocative efficiency in traditional agriculture fails. In order to support his hypothesis, Schultz has taken help of two empirical studies in two villages, one each in Guatemala and India.

The village in Guatemala tends to be capitalistic in character. There is perfect competition in the market and all factors at the disposal of the community are fully used. "All the evidence revealed in the careful documentation of the behaviour of the people in penny capitalism and in the many tables showing prices, costs and returns strongly support the inference that the people are remarkably efficient in allocating the factors at their disposal in current production. There are no significant indivisibilities in methods of production, none in factors, and none in products. There is no disguised unemployment, no underemployment Regarding the village in India, similar results have been drawn by a researcher and widely quoted in Schultz's book.

From the data of this village, the researcher infers that, "there is a remarkably close correspondence between the various price estimates. It would appear that the average allocations made by the sample of farms were efficient within the context of the prevailing technical relationships. There is no evidence that an improvement in economic output could be obtained by altering the present allocations as long as the village relies on traditional resources and technology."

Zero Marginal Productivity

There have been several significant contributions in development economics on the theme of zero marginal productivity. The spurt to this type of exercise was given by

W. Arthur Lewis in 1954. These contributions have developed the theme, at length, that in traditional agriculture, there is a sizeable proportion of labour force who remain disguisedly unemployed and make no contribution to the total product. The implication of this hypothesis is that if a portion of labour force is siphoned off from the farm sector and used elsewhere, the total output of agricultural sector will not fall. This large-scale surplus labour has been put forward as a characteristic feature of traditional agriculture contributing to its poverty. Schultz refutes this hypothesis as well.

He used 1918-19 influenza year as a test of this hypothesis. Influenza epidemic during the year took a heavy toll of life in rural India. He observed that in the crop season that immediately followed the epidemic, the total output of agriculture declined and, therefore, inferred that there could not be even a 5 per cent labour force that could be declared as surplus.

High Returns to Low Capital

Another deep-rooted notion of traditional agriculture is that there is relatively low capital stock and accordingly high rate of return on capital. European countries diverted their investments to poor countries during the colonial period. It also happened that rich peasants of these poor countries invested in western developed countries. Schultz does not accept this notion. Quoting from his two illustrations he shows that there is relatively a large capital in traditional agriculture and hence a low rate of return. He finds that imported capital in the poor countries was not used in building up the traditional factors of production but in transport, etc. Schultz concludes that, "rent from land is a small part of total factor costs in some poor agricultural communities.

This conclusion is quite compatible with economic theory, although the possibility that rent can be small or even zero is sometimes overlooked in considering the contribution of the several factors to agricultural production. The second conclusion is that reproducible material capital is relatively large as a factor of production in many poor agricultural communities, it will not be easy to make this conclusion feel at house that economists have built. To make it even more of a stranger is the fact that the factor share of reproducible capital is large in some of these communities even though the rate on such capital is low.

If the rates of return were always high, the stock could be small and the share relatively large, but if the rate of return is low, as it is in many of these poor agricultural communities, the matter would indeed be baffling given the conventional economic conception that the stock of reproducible capital in poor agricultural communities, is small."

Schultz equally refutes the contention that large-size farms are more efficient. There is no correlation, according to Schultz, between the size of farms and productivity. Small and large-size farms may be equally efficient or inefficient in different situations. Marx, in particular, developed the theme of efficiency of large farms and talked of the returns to scale. Schultz attributes this to political considerations.

Development of Traditional Agriculture

Traditional agriculture, according to Schultz, can be transformed into a relatively cheap source of economic growth. Schultz asserts that transformation of agriculture predominantly depends upon the availability and price of non-traditional (modern) agricultural inputs. He says, that "the supplies of these factors in a very real sense hold the key to such growth." Producing and distributing these factors cheaply makes investment in agriculture profitable and farmers accept these modern inputs and learn how best to use them. This also stimulates the savings and builds up institutions to finance investment in agriculture. Schultz firmly believes that the supplies of modern factors have not been given due attention by economists. They are the producers of the factors of production concealed under "technological change". Some of them engage in research and some in developmental activities. Some produce only information.

The principal sources of the high productivity of modern agriculture are reproducible sources. These sources are of two parts: *(i)* modern material inputs and *(ii)* farm people with modern skills.

These material inputs are selected in the form in which these can be tried in poor peasant economies. Their adaptability to these shall have to be worked out by making several adjustments in these resources and by giving them the form in which these can be used. Research and development plays a significant role in this regard. Availability of modern material inputs elsewhere provides useful information and knowledge Which goes into making the research fruitful in developing adaptable and acceptable varieties of these inputs for poor countries.

In order to produce these inputs, scientific and technological knowledge has to be built up. It is also essential to socialise basic and developmental research. To produce these inputs what one requires are: an optimum size of investment and operational unit; a large number of competent scientists and assistants and adequate amenities for experimental work.

Distribution aspects of these inputs are equally important. The distribution may be done by profit-making firms or non-profit agencies. The profit of the distribution firms depends on the costs of entry and the size of the market. Since costs of entry are high and the market is small, there is not a good scope for profits. In order, therefore, to attract private firms, distribution of these inputs has got to made profitable.

Another crucial factor is provision of information about the distribution of these inputs. Countries like America have transformed and modernised agriculture through these inputs but in poor countries like Guatemala, these services are completely missing. Political conditions and policies also come in the way. These conditions become strenuous if a foreign private firm is trying to enter the market.

Non-profit agencies can also play a role in supplying these inputs – poor countries of the world may specially design such services. Local experiment centres may be set up and farmers induced to test new varieties of seeds under local conditions. A network of agricultural extension services has to be set up.

Farmers as Demanders of New Factors

> "According to urban Folklore, agriculture is the Gibraltar of traditionalism; hence to think of farmers forsaking custom and demanding new factors of production is absurd. This myth created by vested urban outlook cannot be accepted because if it were so, there would not have been modern agriculture anywhere in the world. Examples of farmers of Denmark, Japan, Mexico and other countries explode this myth. The rate and pattern of demand may not be unique but it has always been there. According to Schultz, "the notion that all farmers are handcuffed by tradition, making it possible for them to modernise agriculture belongs to the realm of myth."

The rate of acceptance of new inputs depends upon the cultural variables, on the one hand and observed difference in the profitability, on the other. In a study on "Hybrid Corn", Griliches remarks that, "one of the major factors accounting for the difference in the rate of acceptance by hybrid corn in different areas was the difference in the absolute profitability of the shift-over from open pollinated to hybrid varieties." That farmers respond favourably to profitability criteria and do not remain inhibited by culture has been shown in other research studies as well.

Realising that acceptance of new factor-inputs shall depend upon the profitability, it is necessary to know the components of determining profitability. These may principally be listed as profit and yield. Both absolute and relative yields have to be considered. The difference in yield from year to year shall also be taken into account.

Search for new factors, learning how to use them and costs and returns play a critical role in the transformation of traditional agriculture.

Finally, Schultz asserts that investment in human capital has radical social and economic implications. In this direction, the role of schooling is emphasised. People learn while young, skills are formed at a relatively young age and new ideas weighed against traditional outlook through better education. Importing skills from abroad have many snags. Schultz wants, peasant economies to learn lessons from industrialisation. The belief that investment in education would shrink investment in ore productive enterprises is out rightly rejected by Schultz. According to him, education is the best form of investment in human capital.

Schultz firmly believes that it is only through the right forms of investment that traditional agriculture can be transformed into a modern business proposition.

Schultz's whole thesis of traditional agriculture may be summarised in his own words:

> "The man who is bound by traditional agriculture cannot produce much food no matter how rich the land. Thrift and work are not enough to overcome the niggardliness of this type of agriculture. To produce an abundance of farm products requires that the farmer has access to and has the skill and knowledge to use what science knows about soils, plants, animals and machines. To command farmers to increase production is doomed to failure even though they have access to knowledge. Instead, an approach *that* provides incentives and rewards to farmers is required. The knowledge that makes the transformation possible is a form of capital, which entails investment. Investment not only in material inputs in which a part of this knowledge is embedded but importantly also investment in farm people."

8

Fixing Price in Agriculture

Price determination is a process of balancing demand with supply. In a capitalistic economic order, price system is an important means of allocating society's scarce resources among alternative used. The phenomenon of pricing is a reflection of scarcity of goods and services in relation to their demand. In a competitive economic system, pricing performs three most important functions: (1) It directs the consumption of products and services. It helps a consumer in deciding what to buy and how much to spend on each product, taking into account the relative prices of the products and services and the amount of satisfaction expected to be derived from them; (2) It also directs the use of resources in production. It helps the farmer-producer in deciding what to produce, how to produce and how much to produce. These decisions are based upon relative prices of products and inputs, along with the information on physical input-output relationships; (3) It helps to distribute income among the owners of productive resources. Price variations result in the redistribution of incomes and property values among owners of productive agents. Thus, we find that the knowledge of pricing is of vital importance to the farmer-producer.

The term "price" may be defined as the value received in exchange of goods and services and distinguished from money which is only a medium of exchange. However, the value of an article is itself an abstract concept and depends upon its marginal utility.

Price Determination under Perfect Competition

We have different market situations prevailing in a society, viz., perfect competition, monopoly, monopolistic competition, oligopoly, etc. But in the agricultural sector, conditions of perfect competition are more prevalent than that of any other market situation. We shall, therefore, confine our discussion of price determination in a simplified market, operating under conditions of "perfect competition". The conditions that exist in a perfectly competitive market are as follows:

1. There are a large number of buyers and sellers so that no individual can affect prices by his own actions.
2. The product produced by all sellers is homogeneous so that there is no basis for preferring one seller's product to another.
3. Both sellers and buyers have perfect knowledge about the prevailing price in the market.
4. There are no legal or institutional restrictions on buyers and sellers.

There should be no difficulty in presuming the existence of the conditions listed above for most of the farm products. For example, farm products are generally produced by numerous farmers spread over the entire country and the produce of any individual farmer is so little that it cannot affect the price prevailing in the market.

Further, in most cases, product differentiation is not possible since farmers in the market, for a specified product handle the same identical product and there is no basis for preferring one seller's product to that of another. On the side of consumers also, this type of market situation could be reasonably assumed. They, too, are price-takers. They, too, as individuals, buy so little of any one commodity that each act of purchase has no appreciable influence on price.

A competition that is not pure is called imperfect competition. It exists in two forms, monopolistic competition and oligopoly. In monopolistic competition, though there are many producers but each produces and sells differentiated products. Oligopoly refers to a few sellers, either producing physically homogeneous products or differentiated products. There is an extreme case also when we have practically one seller. It happens in the case of monopoly. All these three market situations practically do not prevail in the agricultural sector.

Within a competitive environment, market prices are determined by the intersection of supply and demand. Demand is a decreasing function of prices while supply is an increasing function. The price is fixed by the point at which the two curves intersect, indicating that demand equals supply. It would be appropriate here to discuss, *a priori*, the demand and supply separately.

The Demand

The demand for a commodity is a schedule of the quantities that buyers would be willing to purchase at different prices at any given time or during a given period of time. The demand for a commodity is the total of the demands of the individual buyers in a market. Two important things must be noticed about the quantity demanded. Firstly, it is a "desired" quantity. It is how much the households actually wish to purchase and not necessarily how much they actually succeed in purchasing.

If sufficient quantity of a commodity is not available in the market, the amount the households wish to purchase may exceed the amount they actually purchase. Secondly, it must be borne in mind that quantity demanded is always a flow. We are concerned not with a single isolated purchase, but with a continuous flow of purchases, and, therefore, demand of a commodity must be expressed as so much per period of time.

Factors, Influencing Demand

Demand for agricultural commodities is influenced by many factors. Four important factors which influence the household demand for agricultural commodities are listed here. These are :

Price of a Commodity: Household's demand for an agricultural commodity is influenced by the price of that commodity. Usually, the higher the price, the lower will be the quantity demanded.

Prices of Related Goods: It is also influenced by the changes in prices of related commodities. In some cases the demand for one commodity will increase as the price of another commodity increases (when the commodities are close substitute), while in other cases, the demand for one commodity may decrease as the price of another commodity increases (when the commodities are complementary).

Tastes, Habits and Fashions: A household demand for agricultural goods is also influenced by the tastes, habits and fashions prevailing in the society at a particular period of time.

Income : Demand for agricultural goods is also influenced by the household income. In most cases, the larger the income, the greater the quantity demanded. But this rule may not always be true in the case of agricultural goods. Since most agricultural goods are necessities of life and their demand is limited by the extent of human stomach, a rise in income of the household may not result in an increase in the demand for such goods. On the other hand, a rise in the income above a particular level may cause a decline in the demand for agricultural goods.

Thus, we find that the amount of an agricultural commodity a household is prepared to purchase is a function of the price of the good in question, the prices of other goods, the household's income and its tastes and habits. This statement may be expressed in symbols by writing down what is called a "Demand Function".

$$Q_n^d = f\,(P_n\,.P_i\,,\ldots\ldots\ldots\ P_{n-1},Y,\ T)$$

Where q is the quantity that the household demands of some commodity, labelled commodity *n*, p_n is the price of this commodity;

P_1 .... P_{n-1} is short-hand notation for the prices of all other commodities;

Y is the household's income and *T* the tastes and habits of the members of the household.

Law of Demand

We observe in our daily life that the higher the price charged for an article, the less of it will be demanded. And the lower its price, the more units people will wish to buy. Thus, there exists at any one time a definite relation between the price of a good such as wheat, rice, etc. and the quantity demanded of that good. We have seen above that the amount demanded is influenced by many factors. While explaining the law of demand, we assume all other factors except price to be constant.

Thus, other things remaining the same, the law of demand states that more will be demanded at a lower price and less when the price is higher. In other words, there is an inverse relationship between the quantity demanded and price. Based on the law of demand, we can construct a demand schedule for an agricultural commodity.

An individual household demand schedule is a list of various quantities of a commodity which an individual household purchases at different alternative prices. The following table shows the quantities of potatoes that would be demanded each week at a number of alternative prices.

	Price per Quintal	Quantity Demanded in Hundred Quintals per Week.
1	80	1.0
2	60	2.0
3	50	3.0
4	40	5.0
5	20	6.0
6	10	8.0

The above table gives the quantities of potatoes demanded for six selected prices but there is a separate quantity that would be demanded at each possible price. It is evident from table that when the price of potatoes is as high as Rs. 80 per quintal, demand for potatoes is as low as 100 quintals. With a fall in price, the demand has shown a corresponding rise. A market demand schedule can well be obtained from this individual household schedule. The market demand schedule can be obtained by adding up all the demand schedules of the individual households in the market.

Demand Curve

Data contained in the previous table when plotted on a graph give the demand curve as shown in the figure below.

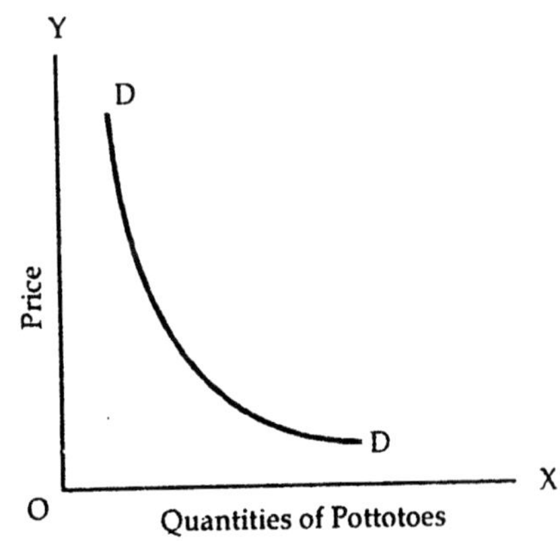

Fig: Demand Curve

DD is the demand curve. This curve shows the quantity of potatoes that the household would like to buy at each price of potatoes.

The downward slope of the demand curve indicates that the quantity demanded increases as the price falls. The curve has been drawn on the assumption that income, tastes and all other prices remain constant.

In the case of inferior goods, the demand curve may be perverse. Inferior goods are those goods which purchasers buy in smaller quantities as they become richer. The demand curve of such goods will only be perverse if they account for a significant proportion of the purchaser's total expenditure. Perverse effects found mostly in the case of staple goods of relatively poor populations since these take a high proportion of their total income, leaving little scope for readjustment of their expenditure elsewhere.

The demand curve that an individual farmer faces for his produce is perfectly elastic. Since he is among the large numbers of producers, he with his own produce cannot influence the market price. In the market, he is a price-taker. At the prevailing market price, he can sell whatever quantity of produce he has produced.

Elasticity of Demand

Consumer's decision as to how much to spend on as commodity at different prices is of great importance to the producers of the commodity. It is so because the responsiveness of the consumers to price changes determines the changes in total revenue resulting from changes in price, and producer's profit is the difference between total revenue and total cost. The law of demand simply shows the direction in which demand will move when the price changes. But it does not show by how much or to what extent a change in price will affect the quantities demanded. With some products, a change in price will bring about a proportionately large increase or decrease in the quantity that can be sold, while with others, the quantity is only slightly responsive to changes in price. This responsiveness of consumers to change in price is known as "elasticity of demand". It is the relative change in the quantity bought resulting from a given relative change in price.

Alfred Marshall was the first economist to give a clear formulation of price elasticity. According to him, it is the ratio of a relative change in quantity to a relative change is price. Let E stand for elasticity. Then:

$$E_p = \frac{\text{relative change in quantity}}{\text{relative change in price}} = \frac{\frac{\Delta Q}{Q}}{\frac{\Delta P}{P}}$$

where Q is quantity, P is price and A(delta) is the symbol meaning "a change in". Thus, *AQ/Q* is a relative change in quantity and *AP/P* is a relative change in price.

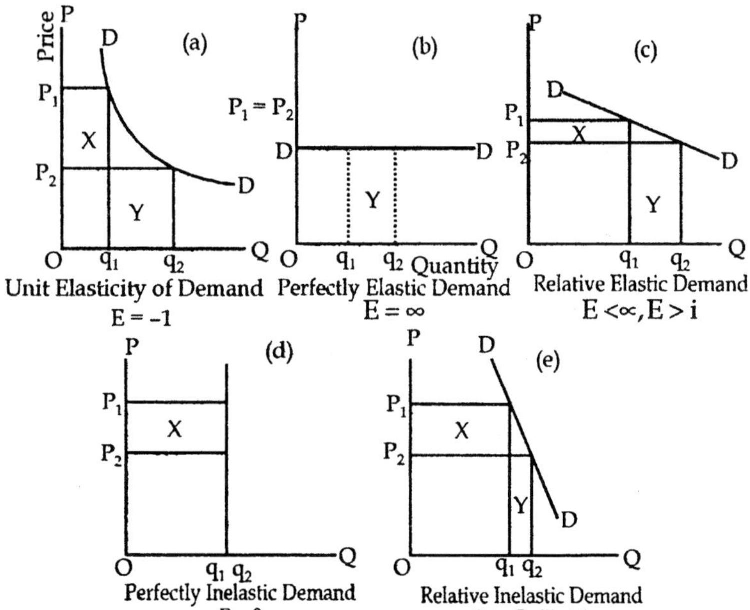

Unit Elasticity of Demand E = -1

Perfectly Elastic Demand E = ∞

Relative Elastic Demand E <∝, E > i

Perfectly Inelastic Demand E = 0

Relative Inelastic Demand E > O, E < 1

There can be five types of elasticities. The numerical value of E may vary from zero to infinity (0 to ∞). If demand is completely unresponsive to the change in price, £ will be equal to zero. In this case, demand is said to be completely inelastic. When the increase in quantity sold is proportionately smaller than the decrease in price, it is a case of a relatively inelastic demand.

In such a case, value of E is greater than zero but less than one $(0<E<1)$. When the proportionate change in the quantity demanded is just equal to the proportionate change in its price, it is case of unit elasticity. The value of £ in this case is equal to one.

The demand is called elastic and the value of £ is greater than 1(£>2) when the proportionate change in the quantity demanded is more than the proportionate change in price. Each of these five cases is illustrated graphically in the figure above (a, b, c, d, e).

The elasticity of demand for individual commodities and for farm products as a whole is very significant in connection with their price determination. The demand for all food products taken as a single commodity is inelastic. The highly inelastic demand for farm products has very important implications for understanding agricultural price policy.

Cross Elasticity of Demand

The relationship between the two goods X and Y can either be substitutive or complementary, or it can even be neutral. In the context of these relationships, the cross elasticity of demand measures the economic relationship between commodities on the demand side. The term cross elasticity of demand may be defined as the ratio of proportionate change in the quantity demanded of commodity X to a given proportionate change in the price of the related commodity Y. The cross elasticity between two goods, X and Y is measured by:

$$E_0 = \frac{\text{The percentage change in the quantity purchased of X}}{\text{The percentage change in the price of Y}}$$

The sign of this fraction tells us whether the two goods are related as substitutes or as complements, and its size tells us how close the relationship is. If X and Y are perfect substitutes for each other, a slight rise in the price of Y will reduce the demand for Y to almost zero. In this case, the cross-elasticity of demand will be infinity.

On the contrary, if two goods are not substitutes at all, the value of cross-elasticity will be zero. Thus, depending upon the degree of substitutability existing between two substitutable goods, the cross-elasticity of demand varies between two extremes – infinity and zero.

In the case of two goods which are complements, instead of being substitutes, the cross-elasticity will be negative. A rise in the price of Y will not only lower the quantity demanded of Y but also a decrease in the quantity demand of X, because both are complementary. A rise in the price of bread might, for instance, cause decreased demand for butter.

Income Elasticity of Demand

Income elasticity of demand for a commodity shows the extent to which a consumer's demand for that commodity changes as a result of a change in his income. It is expressed as:

$$E_1 = \frac{\text{The percentage change in the quantity demanded}}{\text{The percentage change in income}}$$

other factors, in particular the price of commodity, remaining unchanged, so that the change in income is the exclusive cause of the change in purchases. The fraction is normally positive, a rise in income inducing some increase in demand, and a fall, some decrease. The income elasticity will be negative only in the case of an inferior good, and since most of the agricultural goods fall under this category, negativeness of income elasticity for farm products plays a significant role in the pricing behaviour of agricultural products.

There is, of course, considerable variation in the income elasticity of demand for different food products. It tends to be highest for the "protective foods" such as fruit, vegetables, eggs and meat, goods which are relatively expensive per calorie and contain a larger amount of minerals, vitamins or proteins. It is lowest for cereals and potatoes which may indeed be inferior goods. Agricultural products which are industrial raw materials, such as wool or rubber, may have a rather high income elasticity.

If the changes in price are very small, the point elasticity of demand method is used for measuring the responsiveness of demand. In case changes in the price are not small, the arc elasticity of demand method is applied.

The point elasticity of demand is defined as the proportionate change in the quantity demanded resulting from a very small proportionate change in price. Symbolically, we may write :

$$ep = \frac{dQ}{Q} / \frac{dp}{p}$$

or

$$ep = \frac{dQ}{dp} . \frac{p}{Q}$$

If demand curve happens to be linear

$$Q = b_0 - b_1 p$$

Its slope $\frac{dQ}{dp} = -b_1$

Substituting the above formula, we obtain

$$ep = -b . \frac{p}{Q}$$

The simplest way of explaining the point method is to consider a linear demand curve. Let the straight line demand curve be extended to meet the two axes as shown below.

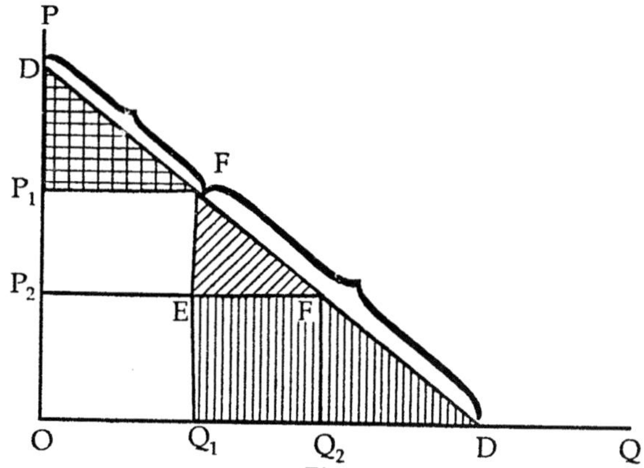

Fig.

We take any point *F* on this demand curve which divides the demand curve into two segments. The point elasticity is, thus, measured by the ratio of the lower segment of the curve below the given point to the upper segment.

The elasticity of the linear demand curve at point F = $\frac{FD'}{FD}$

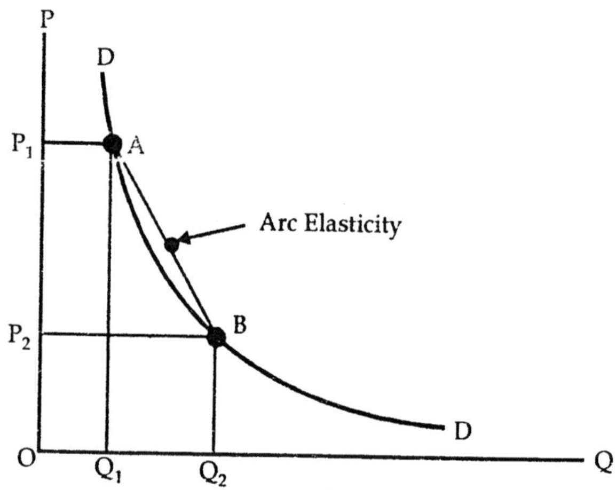

Fig.

The above formula is used to measure the elasticity of demand only for infinitely small changes in price. If price changes are appreciable, we use the following formula, which measures the arc elasticity of demand.

$$ep = \frac{\Delta Q}{\Delta P} \cdot \frac{\frac{P_1 + P_2}{2}}{\frac{Q_1 + Q_2}{2}} = \frac{\Delta Q(P_1 + P_2)}{\Delta P(Q_1 + Q_2)}$$

In the measurement of arc elasticity, we use the average of the two price figures *A* and B as shown in Fig. above (Original and subsequent).

However, this method gives an approximation of the true elasticity of the section *AB* of the demand curve. Clearly, the more convex to the origin the demand curve is, the poorer the linear approximation attained by the arc elasticity formula.

Supply in Practice

By the supply of a commodity we mean the amount of that commodity that producers are able and willing to offer for sale at a given price. Supply may be carefully distinguished from stock and production. Stock constitutes the potential supply but supply at a time is that part of the total produce which the farmer is willing to sell is the stock and is known as marketable surplus and the portion which is actually brought to the market at a particular time for sale is the supply of the produce and is labelled as marketed surplus. Similarly, supply of the farm produce, sometimes, may be more than the current production if past stocks are also brought to the market along with it.

Factors, Influencing Supply

Factors which influence the supply of agricultural commodities are numerous. Given below is a brief introduction to these factors:

1. The supply of an agricultural commodity would depend upon its cost of production, that is, the prices of factors of production which are involved in the production of said commodity. For example, a rise in the price of land will have a large effect on the cost of producing wheat. Thus, a rise in the price of factors of production will

cause the supply of the product to decline and a fall in the prices of factors production may lead to an increase in supply.

2. Supply of an agricultural commodity is also affected by the price that commodity commands in the market. Other things remaining the same, the higher the price of the commodity the more profitable will be to increase its supply.
3. The supply of an individual agricultural commodity will be affected also by the prices of other agricultural goods. An increase in the price of other commodities will make the production of the commodity whose price has not risen relatively less attractive than it was previously. This mainly leads to changes in cropping pattern in agriculture.
4. Supply of agricultural goods also depends upon the state of technology. Agricultural technology helps in bringing down the cost of production and hence increases the supply.
5. There are a number of other factors which affect the supply of agricultural goods, viz. adequate and well spread-out rainfall, improvement in irrigation facilities, increased supply of chemical fertilizers and manure and better and improved methods of production.

We may summarise the above discussion as follows: the supply of an agricultural commodity is a function of the price of that commodity, the prices of all other agricultural goods, the prices of the factors of production, technology and the natural factors. This statement may be expressed in symbols by writing down what is called a supply function:

$$q_n^s = S\left(P_n, P_i....P_{n-1}, F_1....F_m, G, T\right)$$

Where $cj_n s$ is the supply of agricultural good, n, p_n is the price of that commodity P_i ....P_{n-1} stand for the prices of all other agricultural goods, F_i .... F_m denote the prices of all factors of production, *T* is the state of technology and G stands for natural factors.

Law of Supply

As already mentioned, supply has a functional relationship with price. In order to know how the supply of an agricultural commodity varies with its own price, we will assume other things (which also affect supply, to be constant), i.e., $q_n^s = S(P_n)$. With this assumption of other things remaining the same, we can define the law of supply as the quantity of a commodity produced and offered for sale will increase as the price of the commodity rises and decrease as the price falls. Based on the law of supply, we can construct an imaginary supply schedule for an agricultural commodity.

A supply schedule represents the relation between prices and the quantities that sellers are willing to sell in the market.

Table given below shows the quantities of potatoes that would be produced and offered for sale each month at a number of alternative prices.

Table: A Supply Schedule for Potatoes

	Price of Potatoes (Rs. per quintal)	*Quantity Supplied (Hundred quintals per month)*
1	20	5.0
2	40	46.0
3	60	77.0
4	80	100.0
5	100	115.0
6	120	122.0

As is evident from the above table, when the price of potatoes is as high as Rs. 120 per quintal, farmers are ready to offer as much as 122 hundred quintals of potatoes for sale. But the amount of potatoes offered for sale decreases as the price falls. Thus, as the price falls, supply is reduced and as the price rises, the supply is extended.

Supply Curve

Data contained in the previous table could be plotted on a graph to get the supply curve as shown in the following figure.

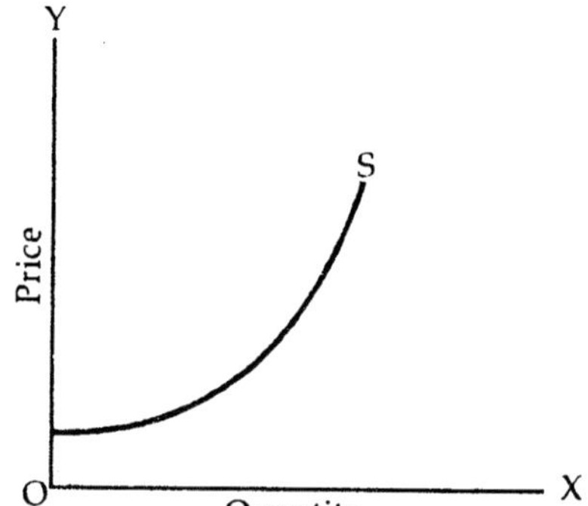

Fig: Supply Curve

Quantity is measured along OX and price along *OY*. *SS* is the supply curve. The curve SS shows the relation between the price of the commodity and the quantity producers wish to sell. It slopes upward indicating that higher the price, the greater the quantity producers will supply. Further, it has been drawn on the assumption of the constancy of all other factors that influence supply other than price of the commodity.

Supply Lags

All supply decisions take time to implement. We may say that supply reacts with a time lag. In cases where the lag is short, it may not matter if it is ignored but in the case of many agricultural commodities, the lag is of critical importance. The simplest possible time lag is one in which this year's price has no effect, whatsoever, on this year's supply. Farmers while deciding about what crop to plant and how much of it to plant look to the existing market price and thus next year's supply depends on this year's price, while this year's supply is based on last year's price. It means:

$$S_t = f(P_{t-1})$$

that is, supply at time period *t* depends on the price of the product ruling in the previous time period.

Perverse Supply Curve

It has often been categorically stated that the supply curve of the produce of the small family farmer will be perverse. The existence of such a supply curve, at least for the peasant farmer in Asia and Africa, was accepted as a dogma between the period of two great wars.

It was based on the assumption that such farmer had a certain, rather low "target" or cash income and that, once this was attained, the farmer would only work as hard as was necessary to maintain it. If the price of his produce rose, he would produce less and enjoy more leisure, if it fell, he would produce more in an attempt to enjoy the same income as before. Studies that have been made to test this hypothesis for commercial crops have not, however, borne it out.

There are two types of perverse response of supply to price. First, supply of farm products may rise in response to price rise to a certain point, but thereafter, it tends to decline though the price continues to rise. This has been shown below:

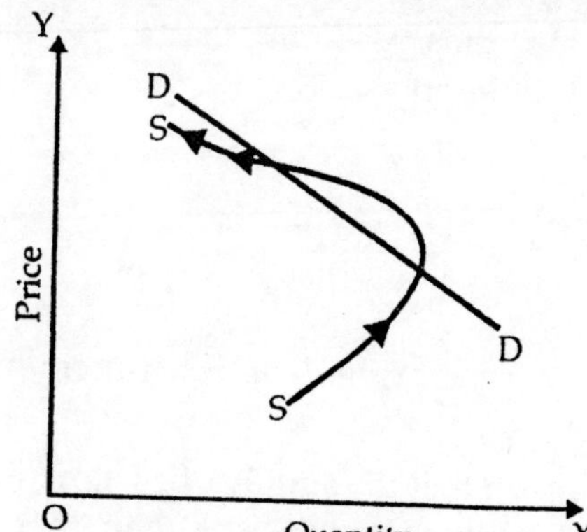

Fig: Poverse Supply Curve.

The situation seems to be rather irrational, if we take into account only the price of the commodity as the sole determinant of its supply. But mostly, the price of inputs used in the process of production, the prices or the cost of production of other agricultural commodities which can be produced as alternatives undergo a change, and given that demand is relatively price inelastic, it is just rational for the producers to reduce production of the first commodity and produce more of other commodities.

Secondly, supply may not fall in response to a falling price. This reaction is typical of agricultural production.

Elasticity of Supply

When the elasticity of supply is referred to with no qualifications, it is taken to mean price elasticity. It measures the direction and extent of the seller's response to a change in the price of his product, and is expressed as:

$$E_s = \frac{\text{The percentage change in the quantity supplied}}{\text{The percentage change in the price}}$$

where the change in price is the exclusive cause of the change in supply. As in the case of demand, we call the supply inelastic when this fraction is less than one, and elastic when it is greater than one.

In continuous terms, assuming that $x = f(p)$, the co-efficient of the elasticity of supply is:

$$\frac{dx}{dp} \times \frac{P_x}{x} = \frac{\frac{dx}{dp}}{\frac{x}{P_x}}$$

= marginal of the function divided by average of the function.

It is well known that the supply of all agricultural produce taken together is much less elastic than that of industrial products. According to a study made by Miss Cohen, during the worldwide depression of 1929-32, total world manufacturing production fell by 37 per cent but agricultural production by only 1 per cent. In the United States, the price index of non-agricultural commodities fell by 38 per cent from 1929 to 1932 while for agricultural produce, it fell by 54 per cent.

During the same period, industrial production fell by 47 per cent, while agricultural production was actually higher in 1932 than in 1929, although by the following year, it was three per cent lower.

Price Determination

Demand and supply can now be brought together in the determination of market price. Let us imagine a wheat market where there are many buyers and sellers who want to buy and to sell. In a free market economy, the prices are determined by the conditions of supply and the demand and the function of the market price is to equate the two. In the short-run, which in relation to agricultural commodities is often called the "stock period", that is, the period after the harvest has arrived in the market, the supply is given.

The market supply schedule shows only the amounts the sellers would sell at different possible prices. The perishability of the commodity, its storage costs, the price it can fetch in other markets if they exist – these matters, together with the sellers' price expectations and their cash positions, determine the shapes of market supply schedules.

In the short-period, market demand schedule is also influenced by buyers' expectations of future prices and availabilities of substitute commodities, together with the price elasticities of their demand and their cash positions.

The equilibrium price in the market period is the price that equates the quantity demanded with the quantity supplied. At this price, the buyers are willing to purchase certain amount of wheat and the sellers are willing to sell exactly the same amount. The market is cleared though some buyers and some sellers remain disappointed since the equilibrium price is too high or too low for them.

Figure given below shows how equilibrium market price is determined in the short period. *DD* and *SS* are the demand and supply curves which have been drawn on the basis of demand and supply schedules just discussed. The two curves intersect at point *A*, which corresponds to the price OP_1 and to the quantity bought and sold P_1A. At this price (OP_1) demand and supply are equal.

The slope of the demand function (D′(P)} is always negative. The slope of supply function *[S′(P)]* is always positive in the absence of external economies. Since the commodity is homogeneous, a single price must prevail. The quantity demanded must equal the quantity supplied at the equilibrium price, so that,

$$D(P) - S(P) = 0 \qquad ....(1)$$

If the price rises to OP_2, the quantity demanded would be P_2B whereas the quantity supplied is P_2C. The excess supply *BC* would force the price down. On the reverse, if price falls to OP_3, we will have EF excess demand, which would force the price up. Thus, OP_1 is the equilibrium price, or the price which balances supply and demand.

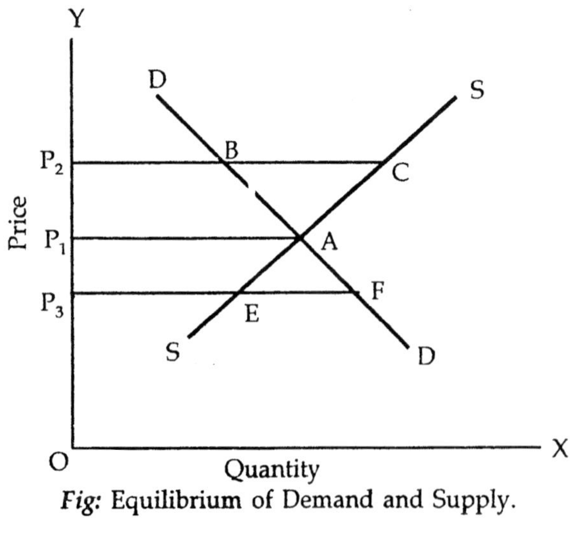

Fig: Equilibrium of Demand and Supply.

When the equilibrium price is reached, neither consumers nor producers have an incentive to recontract, any further. At equilibrium price OP_1' both buyers' and sellers' desires are satisfied and no one can benefit from further re-contracting.

The equilibrium price-quantity combination must satisfy both the demand and supply functions. The equilibrium price is determined by solving the equilibrium conditions 1 for *P*. The equilibrium quantity is determined by substituting the equilibrium price in either the demand or the supply function. Since the equilibrium price-quantity combination satisfies both the demand curve and the supply curve, the above operation is equivalent to finding the coordinates of the intersection point of the demand and supply curves.

This model does in fact reproduce conditions in some agricultural markets at single points of time or over very short time periods. But there are imperfections in agricultural markets and peculiarities in the supply of agricultural products which make the model of only limited use.

Changes in Demand and Supply

Price changes are brought about by shifts in either demand or supply curves or both. If supply remains unchanged while demand increases, a higher equilibrium price is called for. Figure given below illustrates a hypothetical situation.

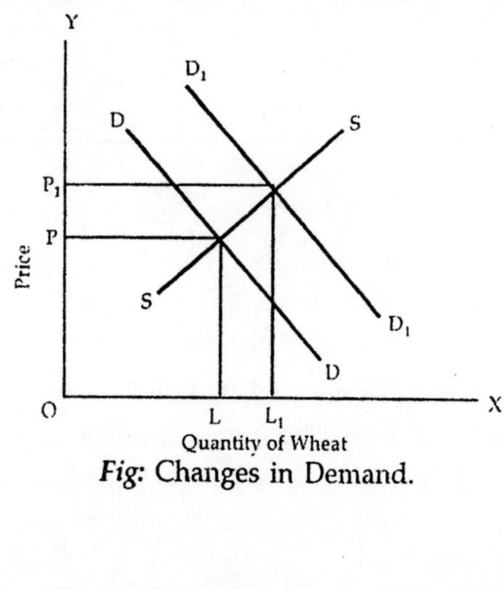

Fig: Changes in Demand.

In above figure, whereas the supply curve based on annual production of wheat in a country X remains unchanged, the demand curve shifts to the right (from *DD* to D_1D_1) indicating an increase in the quantity of wheat that could be sold at any given price. Under original market conditions, the price would have been *OP* per quintal and the quantity supplied would have been *OL* quintal. Due to a shift in demand from

DD to D_1D_1 the new equilibrium is at the price OP_1 per quintal of wheat which corresponds to the supply of OL_1.

In some cases, price changes may be initiated by changes in supply while the demand curve remains in the same position. Figure given below illustrates the effect of shifts in the supply curve for wheat.

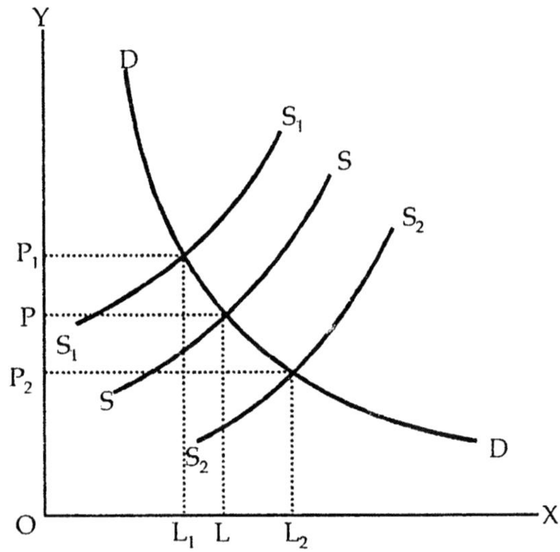

Fig: Changes in Supply

In above figure when the supply curve shifts to the left (decreases), the price rises from *OP* to OP_1 and when it shifts to the right (increases), the price falls to OP_2.

In case both the supply and demand curves change their positions, the direction and theoretical amount of price changes are less certain than in the cases just described.

Exceptions in Agriculture

The response of supply as well as of demand to price changes in agriculture may not exactly be the same as visualised by the law of demand and supply. There is one very important difference between the demand for agriculture and for industrial products. Agriculture, as we know, produces mainly foodstuffs, which are one of the main necessities of life. As a result, the demand for all agricultural products, taken together, tends to be inelastic.

The main reason for this is that capacity of the human stomach is limited. When people improve their standard of living, they substitute better food for plain fare. But beyond a certain point, when their incomes increase, people start spending more on non-agricultural goods. Similarly, when prices of agricultural goods fall, consumption will not greatly expand, nor will it greatly contract if food prices rise.

Accordingly, a low price elasticity indicates that consumers are not sensitive to changes in prices of food on the farm level. This insensitivity stems from two reasons: (1) food is necessary for life, thus changes in its price do not affect its quantity demanded drastically, and (2) with a rise in the standard of living, people spend relatively a small fraction of their income on food which reduces the sensitivity of consumers to changes in the price of food. Econometric studies have shown that price elasticity with respect to food on the farms level is in the order of magnitude of 0.20 to 0.25 which means that if the demand curve for food does not shift, then a 10 per cent increase in the quantity of food will be cleared by the market only after the price of food is cut down 40 per cent.

Because of the inelastic demand for farm products, it may be said that at any particular time any factor which has the effect of increasing the quantity of agricultural products will also have the effect of reducing gross receipts. Any factor limiting the quantity marketed will have the effect of increasing gross receipts of farmers above what they would otherwise have been.

Similarly, aggregate agricultural production is not very responsive to price changes. It is mostly assumed that the farmer is less responsive to price changes. In most cases, agriculture tends to use much the same amount of the factors of production especially land and labour, even when the general level of prices is falling. In consequence, the level of total output in agriculture remains remarkably stable from year to year, notwithstanding the fluctuation in the general price level. The following are the main factors which help to explain the farmer's relative insensitivity to price changes.

The Cost Structure of Agriculture: Costs incurred, whether by the farmer or by the manufacturer, may be divided into two broad categories as escapable and inescapable. Escapable costs are those which, in a given time period, may be avoided by reducing production; inescapable costs are those which, in the same time period, have to be met if the entrepreneur is to stay in business at all. In agriculture, the proportion of inescapable to escapable costs is much higher than in industry where the greater part of manufacturing costs is made up of variable costs. This means that, in bad times, there is less scope for the farmer to reduce his expenses by reducing his production, while every unit of production he can sell will help towards covering his high inescapable costs.

The Small Scale of Farming Business: The great majority of farmers the world over are family concerns, employing little hired labour. In bad times, he may consider a very large fall in family income preferable to leaving agriculture altogether. This is an important factor in the inelasticity of total agricultural supply in the face of a price fall.

Time Element: Another factor which has an influence on the response of farmers to changes in the relative prices of farm commodities is the time element involved in switching from one time of production to another. There are two considerations. The first is that a long period required to make the change may be a deterrent to making any change at all. The second is that after a change has actually been made, it may still take a long time for it to become effective in adding or substracting supplies.

Subsistence Farming: The most important reason responsible for the insensitivity of the farmer to price changes is that most farmers the world over produce for their own consumption and not for market and are, therefore, not affected at all by price changes. Further, since the scope for large incomes is smaller in farming than in industry, the average level of intelligence is probably also lower; as a result, traditional methods of production are ever more important in agriculture than in industry and only a few of the most enterprising farmers really adapt their output as rapidly as it would be most profitable for them to do.

Role of Nature: Agriculture is a biological process and even if the farmer increases certain acreage of a particular crop in response to price changes, he will not be sure to increase its supply. The total output of many crops varies more with the yield per acre which the farmer cannot control than with the number of acres from which he can.

It does not follow from this, however, that the supply of individual agricultural commodities is not sensitive to changes in individual commodity prices. On the contrary, there is evidence that changes in relative prices within agriculture induce changes in supply reasonable efficiently.

The Cob-Web

We now introduce an elementary dynamic theory of price determination of agricultural products. In this theory, we assume that farmers' output plans are fulfilled but with a time lag and we try to show how planned changes in supply can give rise; to oscillations in market behaviour. Agricultural markets subject to simple one-year time lag are illustrated below.

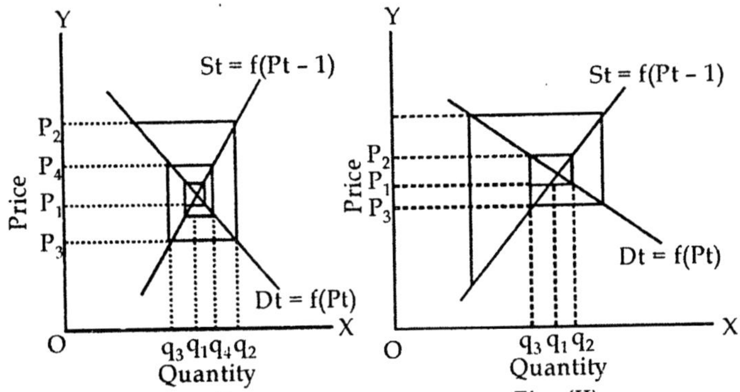

Fig. (I) *Fig. (II)*

In Fig. (I), if in one year *t*, the price is P_2, farmers will plan to produce q_2 in the following year. In the year t_{+1}, q_2 will come in the market and in order to sell this quantity, price must fall to P_3. This level of price (P_3) will induce the farmers to produce only q_3. In the following year t_{+2}, q_3 quantity will be sold at a price P_4. This price in turn will call forth a supply of q_4 the next year, year t_{+3} and this will depress the price below P_4. It is thus clear from the figure that if nothing further disturbs the market the price and quantity will oscillate around their equilibrium values.

In other figure, exactly the same argument as in the previous paragraph applies but in this case, the oscillations become larger and larger so that the equilibrium is never restored.

Thus, while market in Fig. (I) has an adjustment mechanism which is stable, market in Fig. (II) has one which is unstable. The difference between the two figures is that while the demand curve in Fig. (I) is flatter due to which an excess demand and supply can be eliminated with only a small price change, in Fig. (II) the supply curve is flatter than the demand curve which causes the quantity supplied respond more to price changes than does the quantity demanded. In this case, when there is excess supply, a large price fall is necessary to call forth the required demand.

Mathematical Representation of Cob-web Model

The Basic Assumptions of cob-web model are:

1. That the current demand (D_t) is a function of current price (P_t)
2. The supply in current period (S_t) is a function of the price in the preceding time period (P_{t-1}), and
3. The condition for equilibrium is satisfied if the current demand (D_t) is equal to current supply (S_t)

The above conditions can be expressed algebraically as:

$$D_t = A - \beta P_t \qquad ...(1)$$

$$S_t = \alpha P_{t-1} - B \qquad ...(2)$$

and

$$D_t = S_t \qquad ...(3)$$

where A and B are respectively the quantity demanded and supplied which are independent of price and β and a are the co-efficients determining the slopes of demand and supply functions respectively.

Substituting (1) and (2) in (3), we have

$$A - \beta P_t = \alpha P_{t-1} - B$$

$$\beta P_t = A + B - \alpha P_{t-1}$$

A+B $(a)_n$

$$P_t = \frac{A+B}{\beta} + \left(-\frac{\alpha}{\beta}\right) P_{t-1} \qquad ...(4)$$

equation (4) is the general equation, on the basis of which we can formulate the relationship for various periods as follows:

$$P_1 = \frac{A+B}{\beta} + \left(-\frac{\alpha}{\beta}\right) P_0$$

$$P_2 = \frac{A+B}{\beta} + \left(-\frac{\alpha}{\beta}\right) P_1$$

Expressing P_2 in relation to P_g

$$P_2 = \frac{A+B}{\beta} + \left(-\frac{\alpha}{\beta}\right)\frac{A+B}{\beta} + \left(-\frac{\alpha}{\beta}\right)P_0$$

$$= \frac{A+B}{\beta} + \left(-\frac{\alpha}{\beta}\right)\frac{A+B}{\beta} + \left(-\frac{\alpha}{\beta}\right)^2 P_2$$

$$= \frac{A+B}{\beta} + \left[1 + \left(\frac{\alpha}{\beta}\right)\right] + \left(-\frac{\alpha}{\beta}\right)^2 P_0$$

$$= \frac{A+B}{\beta} \times \frac{\beta+\alpha}{\beta} + \left(-\frac{\alpha}{\beta}\right)^2 P_0$$

$$= \frac{A+B}{\beta} \times \frac{\beta-\alpha}{\beta} \times \frac{\beta+\alpha}{\beta} + \left(-\frac{\alpha}{\beta}\right)^2 P_0$$

$$= \frac{A+B}{\beta+\alpha}\left(\frac{\beta-\alpha}{\beta} \times \frac{\beta+\alpha}{\beta}\right) + \left(-\frac{\alpha}{\beta}\right)^2 P_0$$

$$= \frac{A+B}{\beta+\alpha}\left[\left(1-\frac{\alpha}{\beta}\right)\left(1+\frac{\alpha}{\beta}\right)\right] + \left(-\frac{\alpha}{\beta}\right)^2 P_0$$

$$= \frac{A+B}{\beta+\alpha}\left[1-\left(-\frac{\alpha}{\beta}\right)^2\right] + \left(-\frac{\alpha}{\beta}\right)^2 P_0$$

The General Equation Concerning P_t in Terms of P_o can now be Written as:

With the help of the above equation, it is possible to explain perpetual convergent and divergent oscillations.

$$P_t = \frac{A+B}{\beta+\alpha}\left[1-\left(\frac{\alpha}{\beta}\right)^2\right] + \left(-\frac{\alpha}{\beta}\right)^2 P_0$$

Price Determination under Imperfect Competition

So far we have discussed one market situation, which is known as pure competition. As already stated, there are, however, other market situations which are briefly described as under:

Monopoly: A monopoly situation in the market exists when (a) there is practically one producer or seller, (b) difficulty of entry to the industry, and (c) when no close substitutes exist for the product. Both (a) and (c) imply a demand curve with finite elasticity.

Oligopoly: It is a market situation in which the number of sellers dealing in a homogeneous or differentiated product is small. In oligopoly, there is complete interdependence among the sellers with regard to their price output policies. Each seller has direct and ascertainable influences upon every other seller in the industry.

Monopolistic Competition: It is a market situation in which there is a large number of producers with free entry into, and exit from, the industry, but in which each producer sells a product which is somewhat different from that sold by his competitors or we can say that products are differentiated. When there is a large number of firms producing differentiated products, each one has a monopoly of its own product, but is subject to the competition of close substitutes. It is thus clear that monopolistic competition involves both the monopoly and competitive elements.

Monopoly, oligopoly and monopolistic competition all come under the general heading of imperfect competition, characterised by individual sellers whose individual demand curves are not perfectly elastic. The degree of control over prices exercised by any firm operating under imperfect competition is limited by the slope of the demand curve for the product of this individual seller.

On the whole, imperfect competition in agricultural marketing probably has considerably less effect than in industrial production and marketing. Product differentiation, for example, can be much more effective in enhancing the prices of well-advertised face cream, the properties of which are generally a mystery to consumers, than in obtaining higher prices for breakfast butter, the quality of which almost any housewife considers herself a good judge. Consumers more familiar with the real qualities and relative values of foods will not fall so easily for sales propaganda in regard to them.

Determination of Most Profitable Price

The principles of determining the most profitable price for the seller, selling under conditions of imperfect competition as found in most of economics textbooks, are stated in terms of per unit prices and quantities. They involve the use of concepts, such as marginal revenue and marginal costs, which are unfamiliar to most farmers. This principle, however, can be simply stated in terms of total revenue and costs, which are more easily understood.

In Fig. below, *DD* shows the demand curve of an individual producer in and imperfect competitive situation. The horizontal axis represents sales in thousand units while the vertical axis represents the price per unit of product in rupees.

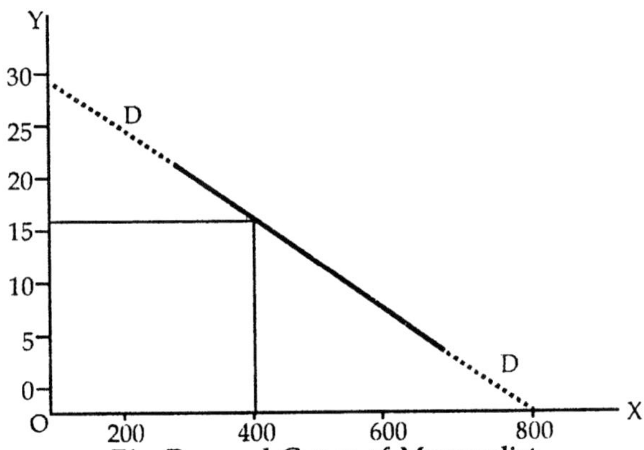

Fig: Demand Curve of Monopolist.

It is evident from above figure that the producer can establish his price in this case anywhere between zero and Rs. 28 per unit. Any price higher than Rs. 28 per unit would bring his sales to zero.

In the following figure, the horizontal scale is the same as in the previous figure representing sales in thousand units. The vertical scale, however, is different. It represents total rupees of revenue derived from the sale of different quantities. The *TR* curve represents the total revenue that would be realised from the sale of each of the quantities specified in the horizontal axis.

The *TR* curve has been derived from the demand curve *DD* in above figure *TC* is the total cost curve, which represents the sum of fixed and variable cost for any given output.

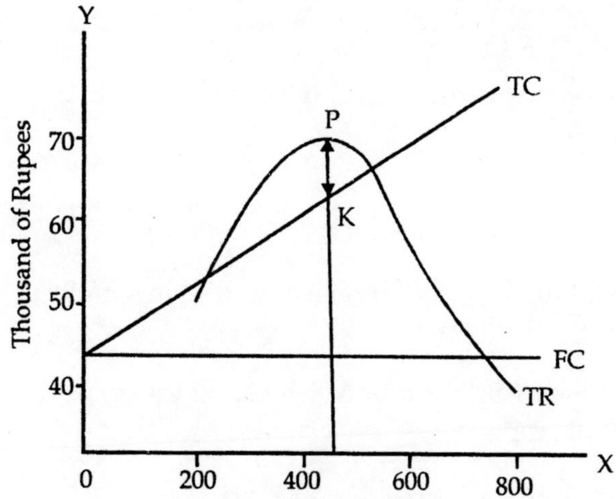

Fig: Profit Maximisation.

The net profit or loss received by the seller by selling any given quantity of the produce is represented by the vertical distance between lines *TR* and *TC*. In case, the line TR is above the line *TC*, a net profit is indicated, since gross revenue is greater than total cost. In this figure, the greatest profit which is reflected by the greatest distance between lines *TC* and *TR* is found at a quantity of 440,000 units. Against this sale, the net profit would be 13,000 (*TR* - *TC*). In order to maximise his revenue, this is the quantity of produce a seller under imperfect competitive conditions would wish to sell.

The monopolist's total revenue and total cost can both be expressed as function of output:

$$R = R(q) \qquad C = C(q)$$

His profit is the difference between his total revenue and total cost: $n = R(q) - C(q)$

To maximise profit, we set the derivative of the above equation with respect to q equal to zero:

dir

$$\frac{d\pi}{dq} = R'(q) - C'(q) = 0$$

or $R'(q) = C(q)$

The second-order condition for profit maximisation requires that

$$\frac{d^2\pi}{d\pi^2} = R''(q) - C''(q) < 0$$

or adding $C''(q)$ to both sides of the inequality,

$$R''(q) < C''(q)$$

The rate of increase of MR must be less than the rate of increase of MC

Price Support

Unplanned fluctuations in supply do occur frequently in agriculture. In those economies where the price of such goods are left to be determined by the free market forces, we experience large price fluctuations. In the case of many agricultural goods, as already stated, the demand is quite inelastic. In these cases, we find very large price fluctuations together with the peculiar situation that when nature is unexpectedly kind and farmers are rewarded with a bumper crop, they see their incomes dwindling, while when nature is moderately unkind and farm supplies fall unexpectedly, farmers' income rises. "Thus, during a year, the market price of foodgrains is determined predominantly by the size of the harvest and the price elasticities of consumer demand for foodgrains. Because of generally low price elasticities of demand for foodgrains, it needs a disproportionately large rise in order to restrict the demand to a somewhat smaller harvest; and it needs a disproportionately large fall in price, to expand the demand to a somewhat large harvest. The market price of foodgrains is thus liable to large fluctuations from year to year, firstly because of fluctuations in the size of the harvest and secondly because of the generally low price elasticities of demand for foodgrains."

On account of the generally low price elasticities of consumer demand for foodgrains, the fluctuations in prices are likely to be more pronounced than warranted by fluctuations in the harvest. Thus, in a year of bad harvest, the rise in price might more than compensate the fall in yield and the producers might actually reap a higher total revenue. On the other hand, in a year of good harvest, the prices might fall so low that in spite of the high yields, the producers might end up with a smaller total revenue. There is, therefore, a strong case for minimum support prices of foodgrains, particularly in a year of good harvest.

From the consumer's point of view also, stability in foodgrains prices is highly desirable. "It does not help the consumer to have a low price and high consumption in a year of good harvest and a high price and low consumption in a year of bad harvest. Therefore, he would prefer the price to be supported in a year of good harvest and thus not be allowed to fall below a certain minimum, provided a revenue obligation is accepted and in year of bad harvest, the price is not allowed to rise above a certain maximum. Thus, in the view of the consumer, a maximum ceiling price is a necessary corollary to a minimum support price."

It is not easy to give a precise definition of "support price". It can be differently defined with reference to the objectives of the price policy in view. Among the three objectives of price policy, viz. stability, equity and growth, growth objective is much more relevant in the present context of many developing countries. Considering this objective, the minimum support price can be defined as a "reserve price" announced in advance of the production period and more or less corresponding to a forward price, which, in conjunction with other measures, will ensure the desired quantities of production and desired relationship between each other."

In simple words, support price may be defined as the price at which the government would be under obligation to buy the entire stocks that may be offered to it for sale. The support price would have no direct incentive role to play, but would insure farmers against the risk of prices falling below a particular level. The support price would also have to be such as would not inhibit adoption of improved technology by progressive farmers.

Under the support price policy, government attempts to stabilise the incomes of farmers by entering the market itself, buying and selling in the open market when there is a shortage.

An important question that must be answered is: "What should be the minimum support price?" Guidelines for determining support prices of agricultural products would depend on the objectives that are sought to be achieved. Objectives of agricultural price support policies can be and are in fact diverse in different countries. By definition a support price policy assures the farmers against a fall in prices beyond the stipulated level. In some countries (mainly advanced), such price insurance has the primary objective of maintaining the general level of farm incomes. In several other countries (especially developing countries), the main objective of support price policy is to help augment overall agricultural production. Thus, on the basis of the primary objectives that they seek to achieve, the support price policies can be categorised as either "Income-oriented" or "Production-oriented".

In most developing countries, including India, the main objective in the present context is to step up the rate of growth of agricultural production so as to match the growth of consumer demand. The support price policy, which is production-oriented, therefore, would seen to have greater relevance. The objective of improving agricultural incomes will be achieved as a sequel to increased production and productivity.

For the fixation of minimum support price, some economists believed that it should be based on the cost of production. But this is not so simple. The main questions that need to be answered are: (a) What costs to be considered? Cost of production of a commodity depends on a number of factors which vary from farm to farm, e.g., size of the farm, soil type, cropping pattern, farm investment as well as technique of production employed. It is also observed that farms in a homogeneous tract do not incur uniform or similar costs. Also standardisation of agricultural produce is not easy as the quality of produce depends mainly on natural factors, among other things. As such it is not easy to work out the average cost of production.

To avoid these difficulties, support price has to be related to the costs of farms for which the inputs are mostly purchased and not home produced. The reserve price has to be related to the cost of production of a "model" commercial farm, for which alone the cost is a measurable concept. The reserve price may thus be defined as "the price that ensures the cost of production of a commercial farm under normal weather conditions." Thus defined, it will vary from year to year, depending upon changes in the conditions of supply, and hence it will have to be flexible.

The cost of production approach, though useful for some purposes, is essentially a backward-looking approach. It cannot ensure the necessary stimulus for increasing production. For this purpose, support prices must have an element of incentive.

The price mechanism is a highly sensitive and delicate instrument and the repercussions of a change in price in one area would be so widespread that it seems almost an impossible task to take into account all of them and arrive at an integrated price structure that may be truly described as in equilibrium, both over space and time. Nevertheless, attempts to collect and analyse as much empirical evidence as possible will have to continue and to be intensified so as to derive supply response co-efficients, which can serve as guidelines for determining support prices. However, we may be sure that the primary objective of the support price policy in our country has to be that of augmenting agricultural production and not of achieving income redistribution between agricultural and non-agricultural sectors. The guidelines for determining support prices will have to be evolved keeping this objective firmly in view.

According to this programme, the government should fix the price of farm produce at a level which is higher than the market price and to buy from the farmers whatever surplus are not cleared in the market. In the following figure, P_0 is the market price of, say, wheat.

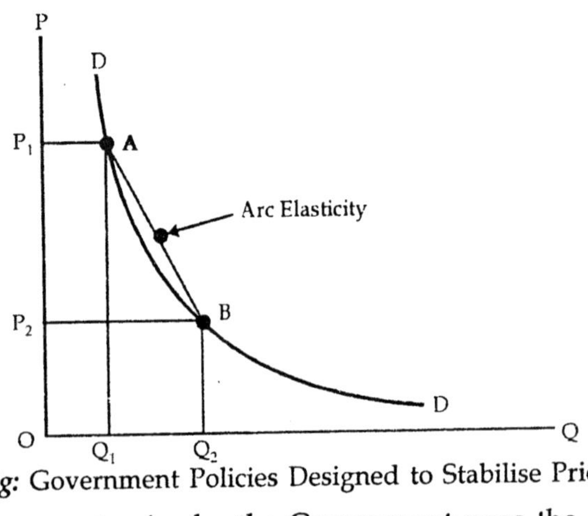

Fig: Government Policies Designed to Stabilise Price.

The support price by the Government pegs the prices at P_1. At that price, according to the demand curve D_0, only OA units of wheat are taken by consumers. But on the other hand, *OB* units of wheat are offered for sales. Thus, the government is under the obligation to purchase this surplus supply *(AB)* and keep it in the buffer stock.

In case this policy is successful, we will, firstly, have smaller fluctuations in the price of farm produce than there would be if price were determined on the basis of a completely free market. Secondly, total revenue of the farm producers will get stabilised in the face of fluctuations in production.

Long-term Considerations

In the short period, since the supply of agricultural produce cannot be increased, the price which the farm operator receives may be determined by the intersection of existing demand and existing supply. Such price determination is more or less independent of the inputs or costs incurred by the farm producers. In the long-run, the size of production is, however, positively related to the inputs or costs incurred. The producer knows that he can expand production by increased inputs or costs. But he would naturally not do it, unless it paid him to do so. Thus, the inputs or the costs that the producer incurs get related to the price that he expects.

Under these conditions, the demand for minimum support price arises on the ground that the producer must be assured of minimum price which he may count upon and which may provide a basis for production decisions regarding inputs and costs to be incurred. Higher support prices fixed by the government may stimulate agricultural production by causing farmers to use more labour and other variable resources inputs to reach higher output levels with existing methods of production, or by inducing investment and the discovery and adoption of new agricultural technologies that result, in new, lower-cost production possibilities by farmers. Thus, a surplus of AB units of output is created in the market. This surplus is purchased by the government at the support price announced by it and kept in the buffer stock.

The purpose of fixing support price, therefore, is to assure the farmer that in case market prices tend to go below then, government will step in and purchase all quantities offered to it for sale at the guaranteed prices. It is obvious that the fixation of support prices does not rule out the possibility of market prices being higher than the support prices. In fact, in a situation where demand is increasing at a faster rate than supply, market price can generally be expected to remains higher than the support price.

Behaviour of Agricultural Prices in India

Agricultural prices, unlike manufactured goods prices, fluctuate more violently as the agricultural sector is highly influenced by the vagaries of nature. The prices of agricultural goods in India have been rising continuously since independence except during the First Five Year Plan, When the prices actually fell.

Table: Trends in Wholesale Prices of Agricultural Goods

Year	*Index Number 1952-53=100*	*Per cent Increase Decrease*
1950-51	110.0	–
1955-56	88.0	–
1956-57	104.5	+ 18.8
1957-58	107.4	+ 2.8
1958-59	114.0	+ 6.1
1959-60	116.5	+ 2.2
1960-61	123.8	+ 6.3
1961-62	115.5	- 6.7
Index Number with Base Year 1961-62=100 Weight 33.2		
1962-63	102.3	+2.3
1963-64	108.4	+6.0
1964-65	130.9	+20.8
1965-66	141.7	+8.3
1966-67	166.6	+17.3

Contd...

Year	*Index Number 1952-53=100*	*Per cent Increase Decrease*
1967-68	188.2	+13.0
1968-69	179.4	-4.7
1969-70	194.8	+8.6
1970-71	201.4	+3.2
Index Number with Base Year 1970-71 = 100 Weight 40.4		
1971-72	100.4	+0.4
1972-73	110.3	+9.9
1973-74	139.2	+26.1
1974-75	169.2	+22.1
1975-76	157.3	-7.4
1976-77	158.4	+0.7
1977-78	174.8	+10.3
1978-79	171.9	-11.7
1979-80	188.7	+9.8
1980-81	210.5	+11.6
1981-82	236.5	+12.3
1982-83	248.3	+5.10
1983-84	283.1	+14.0
1984-85	303.2	+7.2
1985-86	309.6	+2.1
1986-87	330.1	+6.6
Series with Base 1981-82=100 Base Year Weight 27.5		
1982-83	107.3	+7.3
1983-84	121.7	+13.1
1984-85	129.2	+6.4
1985-86	129.1	-0.1
1986-87	142.8	+10.6
1987-88	161.8	+13.3
1988-89	170.9	+5.6

Contd...

Year	*Index Number 1952-53=100*	*Per cent Increase Decrease*
1989-90	174.4	+2.0
1990-91	198.3	+13.7
1991-92	236.7	+19.4
1992-93	255.6	+8.0
1993-94	271.2	+6.1
1994-95	307.7	+13.5
1995-96"	330.5	+7.4
1996-97	358.4	+8.4

Index numbers presented in the above tables make it distinctly clear that there occur sharp variations in the agricultural prices. Since the early sixties, these prices have not only been rising over the years but have also fluctuated at times. The decade of fifties may be treated as the period of stable agricultural prices except for the year 1956-57 which was a very bad agricultural year. In the sixties, prices maintained an upward swing and almost doubled over the decade with some years showing marked variations. During the seventies and late eighties, agricultural prices rose at an alarming rate and more than trebled between the years 1971-72 and 1986-87. In some years, prices went up by as much as 26 per cent.

Agricultural Price Policy

Need for Agricultural Price Policy : As stated earlier, agricultural prices fluctuate violently and frequently under the free market mechanism to the disadvantage of both the producers as well as the consumers. These fluctuations in the prices of agricultural products are the greatest hurdle in the way of agricultural development, for they bring ruin to many. For instance, in the bumper crop years, prices fall too low to leave farmers any appreciable margin and in the light crop years, prices rise so steeply that the farmers have a very little marketable surplus. Similarly, persistent imbalance between supply and demand causes violent fluctuations in consumer prices and thus affects the poor consumers. The two major aspects of the price policy, therefore, are: (i) to protect the farmers interests by removing or at least mitigating the major uncertainties by assuring them remunerative prices for their produce, and (ii) to safeguard the interests of the low income consumers by assuring minimum supplies of food articles at reasonable prices.

Objectives of Price Policy

The major objective of the agricultural price policy is to achieve price stability without destabilising total revenue of the farmer and provide a price support which would be economic to the grower as well as agro-based industry and at the same time, subserve the interests of the consumer. In other words, the intention is to integrate support prices with policies to stabilise prices and supplies to consumers. In the developed countries like the USA, Canada, and Western Europe, where the farm incomes have badly lagged behind non-farm incomes in the process of economic growth, the main objective of price policy is to raise farm incomes so as to bring them in line with the income levels in the rest of the country.

In the underdeveloped countries, however, the income-oriented price policy of the developed countries has not much direct relevance. In poor countries, the problem is not overproduction! In these countries, the objective of farm price policy, therefore, should be to increase agricultural production by creating economic incentives for farmers. The policy should be able to perform the following functions.

1. To accelerate the growth of agricultural output as a whole.
2. To stabilise prices in order to prevent fluctuations.
3. To bring about desired changes in the product mix.
4. To increase the marketable surplus.

5. To ensure adequate supplies of foodgrains to the low income consumers at reasonable prices.

Thus, the "price policy must ensure that agricultural production is economic both in the widest and in the strictest sense of the term. In its narrow sense, economic production would mean that costs are reduced to the minimum, that the agriculturists have a fair margin of profits and that the costs of agricultural products, foodgrains and raw materials as they enter into the costs of living and the prices of manufactured articles either in the internal or external markets are healthy levels.

In the wider sense, economic production would signify the widest distribution of scarce land resources among the various competing ends, forestry, pastures and cultivation in the first instance, and secondly, between food crops and cash crops".

Enunciation of Agricultural Price Policy in India

During the last three decades, India's agricultural price policy has moved through two distinct phases. Upto 1965, the government was following an ad-hoc type policy, marked by spells of hectic activity in years of poor crops and complete complacency in years of good crops. In 1965, however, the Agricultural Prices Commission was appointed and therefore, the subsequent period witnessed the beginning of a more stable and meaningful price policy.

Before 1965

Prior to the beginning of the First Five Year Plan, two committees were appointed by the Government of India who in their recommendations hinted towards the desired price policy. In 1947, the Foodgrains Policy Committee was formed which recommended a policy of progressive decontrol, reduction of dependence on imports of foodgrains and a substantial increase in domestic production within the earliest possible time. Another committee, known as Foodgrains

Procurement Committee (1950) recommended the continuation and extension of the system of rationing in the country.

During the First Five Year Plan period, due to the relatively easy situation on the food front, the government followed a policy of complete decontrol. However, prices of agricultural products started showing a rising trend again in the last leg of the First Plan, forcing the government to introduct again partial controls. Concerned about the rising prices, the government appointed the Foodgrains Enquiry Committee in 1957. The Committee strongly recommended the setting up of a Foodgrains Stabilisation Organisation to stabilise the food price.

It also recommended the building up of a buffer stock, licensing of wholesale traders and fixation of minimum and maximum prices. Thus, till 1964, the agricultural price policy was more concerned with the stabilisation of consumer prices and not much attention was paid to providing any incentive to the producers.

After 1965

It was during 1964 that the government envisaged a price policy which ensured incentive prices to the farmers. The seeds of this change were to be found in the Third Five Year Plan document, which stated that "the producer of foodgrains must get a reasonable return. The farmer, in other words, should be assured that the prices of foodgrains and the other commodities that he produces will not be allowed to fall below a reasonable minimum."

In order to provide fair and economical prices to the farmers, a Foodgrains Prices Committee was appointed in 1964. Its main recommendations were: *(i)* introduction of rationing in big cities; *(ii)* opening of fair price shops in other areas; *(iii)* gradually withdrawing the restrictions on interstate movement of foodgrains; *(iv)* licensing the wholesale trade in foodgrains; and *(v)* strengthening the state administrative machinery. Its main recommendation was the setting up of an Agricultural Price Commission so that "the price policy of all agricultural commodities should come within the purview of the Agricultural Price Commission so that a balanced and integrated price structure could be evolved and the claims of competing crops on limited resource can be resolved in the perspective of the overall needs of the economy."

It was in 1965 that the Agricultural Price Commission (APC) was set up with a view to evolve a balanced and integrated price structure in the perspective of overall needs of the economy and with due regard to the interest of the producer and the consumer. The major functions of APC are as follows:

1. To advise the government on the price policy of agricultural commodities, particularly paddy, wheat, jowar, bajra, maize, gram and pulses.
2. To recommend measures to make the price policy effective.
3. To suggest measures to reduce costs of marketing and recommend fair price margins for different stages of marketing.
4. To advise the government on any problem relating to agricultural prices and production.

The APC does not follow a mechanical approach in deciding upon the price policy. While recommending a price policy for a commodity the Commission takes a comprehensive overview of the entire structure of the economy of that commodity, its demand and supply situation, available date on costs of cultivation, export potential, price trends and the general economic health of the economy Besides, the Commission also takes into account the level of the administered prices for competing crops, so that a measure of inter-crop price parity is achieved. While recommending the price policy, the Commission also suggests such non-price measures as would facilitate the achievement of the objectives of the policy."

Along with the setting up of APC, the government took another step of long-term consequences by setting up of Food Corporation of India in 1965. The main functions of this Corporation are to procure, store and distribute foodgrains.

For providing incentives for increased production, the Fourth Five Year Plan recommended a minimum price for the main agricultural commodities. However, the effectiveness of this policy, according to the plan document, depended on the adequacy of the relevant machinery for purchase, including the Food Corporation of India, State Trading Corporation and the Cooperative Marketing Organisations.

During the Fifth Plan period, two main considerations were kept in view while recommending the agricultural price policy. First, it should provide incentives for sustained and higher production, and second, a discriminating manipulation of inter-crop price relationship to induce the farmers plan the production of different crops in line with the estimated demand.

The Fifth Plan sought to maintain a clear-cut distinction between minimum Support Price (SP) and Procurement Prices (PP). The minimum support price based on the cost of production and other relevant factors is the minimum floor price which would be announced for all important foodgrains crops prior to the sowing season.

The procurement price, on the other hand, to be announced later, will usually be fixed in terms of a premium over and above the minimum support price. The procurement prices thus fixed will naturally take into account the size of the anticipated crop in addition to the requirements of public distribution and buffer stocking.

The Sixth Plan regarded price policy as being of crucial importance in agricultural development on account of the following reasons.

1. Modern agriculture increasingly involves the use of costly inputs as part of improved technology and hence

an assured minimum price becomes a necessary underpinning for sustained agricultural production.

2. Price policy is also an important tool for facilitating crop planning, an aspect which has so far not received adequate attention in the country.
3. Price policy can be geared towards ensuring that the relevant income levels of the farming community are not eroded by continuing unfavourable terms of trade between the agricultural sector and the non-agricultural sector.

Actual Price Policy

The present price policy for agricultural products is guided by the objectives of providing remunerative prices to the producers, to facilitate procurement and also to maintain the requisite inter-crop parity. Under this policy, government has been providing minimum support prices of important food crops and have been making upward revisions in the support/ procurement prices of these agricultural commodities. At the declared support or procurement price, the government stands to purchase whatever amount of that crop is offered for sale in the market.

The procurement price for different crops is determined and suggested to the government by the APC. Government, after due consideration, declares the support or procurement price of a particular food crop before harvesting begins. As a result of following such a policy, the wheat price during 1984-85 was raised by 32.2 per cent, of paddy, by 44.2 per cent and of coarse grains by 36.8 per cent. In respect of two important cereals, viz.; rice and wheat, this policy has yielded rich dividends. Buffer stocks are built up through procurement at support prices which sustain the public distribution system. Over 15 million tonnes of foodgrains are procured by public agencies annually. Enforcement of this policy over the years has demonstrated that agricultural development is directly related to the price policy through which the farmers are assured of a minimum support price.

Important items of foodgrain are distributed among the consumers through ration and fair price shops throughout the country. If the consumers' requirement exceeds the rationed quantity of an article, these can be purchased from the open market at market rates.

Though the thrust of the policy had been to achieve the twin objectives of assuring remunerative prices to the farmer and providing foodgrains to the consumer at reasonable prices, it was amply made clear that the concern for consumer interest should not be allowed to take away the farmer's incentive to adopt improved technology and make necessary investment for the purpose.

The framework of the policy was modified in 1980. The emphasis of the policy, as reflected in the revised terms of reference of APC which was later renamed as Commission for Agricultural Costs and Prices (CACP), shifted from maximising the production to developing a production pattern consistent with the overall needs of the economy. The policy was further reviewed in 1986 when a long-term perspective for agricultural price policy was presented to the Parliament.

The price policy was again subjected to a rigorous review after a programme of economic reforms was launched in 1991. "The package of reforms in agriculture is based on the diagnosis that while the sector remained net disprotected, the subsidies arising out of inappropriate pricing of inputs and outputs led to inefficient resource use, eroded the capacity of the Government to finance public investment in agriculture and benefited only the producers of few crops and that too in some regions.

The suggested agenda for the agricultural sector, therefore, revolves on setting the prices right and includes withdrawal of subsidies on inputs, targeting the public distribution system only to poor, abolition of food management system and its attendant costs and liberalisation of trade in agricultural commodities" Under new economic regime the subsidies on farm inputs and food are no more sustainable in terms of fiscal management and these be phased out and adjustment in agricultural prices be made for arresting the deterioration in terms of trade for the agricultural sector.

At present 24 commodities are covered under the minimum price support programme. These include paddy, wheat, jowar, bajra, maize, barley, gram, tur, moong, urad, groundnut, rapeseed, mustard, toria, soyabean, sunflower seed, sesamum, copra, cotton, Jute-mesta, Virginia flue cured tobacco and sugarcane which together account for 2 per cent of the total value of crop output in the country. These apart, some other commodities like onion, potato, ginger, chillies black pepper, caster seed and some fruits are included under the Market Intervention Scheme (MIS).

In case of cereals, till 1970-71, government used to announce, apart from the minimum support price, procurement prices at which public agencies procured specified quantities of grains from the market. The procurement prices used to be higher than the minimum support prices but lower than market prices. In 1970-71, Government decided to announce only the procurement prices and provide support to the farmers at these prices. However, recognising that the blurred distinction between the procurement prices and minimum support prices had started depriving the farmers of the guarantee that was inherent in the fixation of minimum support price, the Government since 1991, on the recommendation of CACP, has decided to fix only the minimum support price in the case of cereals also.

Evaluation of Agricultural Price Policy

The Price Policy, whether of agricultural products or of manufactures, is formulated keeping in view some objectives.

The objective of stability is, perhaps, the most important one, more so, in agricultural where due to strong natural factors, serious fluctuations can occur. Drastic and frequent fluctuations are deterrent to increased production and result in considerable uncertainty. Government's agricultural price policy is designed by the Agricultural Price Commission (APC) which after taking note of the cost of production fixes procurement prices and minimum support prices of different agricultural commodities.

Procurement prices are those at which the government procures surplus grain from the farmers and support prices are those at which government is bound to purchase the surplus grain if the prices fall below the minimum level. However, the effectiveness of these policies is determined by the level of implementation. Many a times, what has been happening is that the level of implementation would remain below expectations and there would arise differences between the actual procurement prices and the effective market prices. These differences, sometimes, have been found to be so substantial that it has affected adversely market arrivals. Such ineffective implementation and regulation of prices even now leads to fluctuations in agricultural prices which are not desirable.

While formulating a price policy, it has to be kept in mind that the benefits percolate to all sections of the farm population. Usually, it has been noticed that the benefits accrue more to the large farmers than to the small ones. Large farmers have better access to inputs and they generate a higher marketable surplus compared to the small farmers who do not enjoy such access to inputs. Even the system of procurement is such that the small farmer's is not reached in his own village. Government operates through private commission agents and does not directly enter the open market. In some cases, it has also been noticed that large farmers hold back their surplus and sell it at a higher price sometimes later in the open market. This benefit does not accrue to small farmers who are in greater need of cash and sell their surplus only through the procurement channel. Strong regulatory measures are, therefore, required in this regard so that incentives are provided to both large and small farmers in an adequate measure.

The agricultural price policy should also provide safeguards to the consumers, and one of the best ways to do this is through a proper public distribution system. As such, a public distribution system should provide for adequate food through fair price shops to meet the requirements of the vulnerable sections of society, should include all the major crops and should reach the depressed sections in the rural areas. It may be noted that our public distribution system flops on all these counts. Stocks are often inadequate, mainly rice and wheat are covered ignoring other inferior grains which form food for the poor and which do not reach all the rural poor population.

Procurement prices have been raised year after year and agriculture forming the dominant sector in the economy, the general price level has been moving up in full sympathy with the trend in procurement prices. As such, these prices tend to be inflationary. It has been observed that higher procurement prices in the previous years have enhanced the holding power of large farmers which in turn becomes a contributing factor to further rise in procurement prices.

While fixing procurement prices, the APC takes cognisance of the cost of production but not of the return to each rupee invested which may often be more than 100 per cent. As such, under the garb of higher cost of production, the procurement prices are forced to be fixed at a higher level than before and made to contribute to inflationary tendencies.

The agricultural price policy of post-independence era seems to have failed to protect the interests of the rural poor such as marginal farmers and landless labourers. Higher procurement prices give a spurt to foodgrain prices and makes things hard for the poor whose bulk of earnings go to the purchase of foodgrains. Since they do not have any marketable surplus so that they could benefit from enhanced procurement prices, they pay through their nose for their very livelihood.

The agricultural price policy has to be reoriented in a meaningful fashion, blending production incentives with consumer safeguards in a more balanced manner.

Farmers and Terms of Trade

Terms of trade simply denote the terms on which two sectors exchange their goods. Farmers' terms of trade refer to the terms at which agricultural products are exchanged for non-agricultural products. The agricultural sector offers foodgrains and other raw materials to the non-agricultural sector and the former in turn, purchases a large number of manufactured goods and agricultural inputs from the non-agricultural sector.

The terms of trade may be favourable or unfavourable to a particular sector depending upon the relative movement of prices in both the sectors. For instance, if, over a period of time, the prices of agricultural products rise at a faster rate than the prices of the manufactured goods, we will describe this situation as the agricultural sector having favourable terms of trade and *vice-versa.* Farmers' terms of trade in a nutshell may be defined as "the ratio of prices received and prices paid by the farmers and may be explained as:

$$\text{Farmers' terms of trade} = \frac{\text{Farm prices}}{\text{Manufactured goods prices}}$$

If the above ratio happens to be greater than 1, the terms of trade would be deemed favourable to the agricultural sector and in case, the ratio is less than 1, farmers would be losing their income considerably.

Movement of Agricultural Prices and Terms of Trade

Table given below shows how the prices of agricultural commodities have moved in recent years in relation to the prices of manufactured goods as well as to the general price index.

Index Numbers of Wholesale Prices

Relative Prices of Manufactures and Agricultural Commodities

Year	*General Index of Wholesale Prices*	*Index of Manufactured Products*	*Index of Agricultural Commodities*	*Prices of Manufactures as Per Cent of Agricultural Commodities*
Base Year				
Weights	*(100.0)*	*(29.0)*	*(46.1)*	–
1950-51	111.8	101.8	110.0	92.5
1955-56	92.5	99.6	88.0	113.2
1956-57	105.3	105.6	104.5	101.1
1957-58	108.4	108.2	107.4	94.8
1958-59	112.9	108.1	114.0	94.8
1959-60	117.1	111.3	116.5	95.5
1960-61	124.9	122.8	123.8	99.2
1961-62	125.1	124.6	122.9	101.4
Base Year				
Weights	*(100.0)*	*(32.30)*	*(33.2)*	–
1962-63	103.8	103.2	102.3	100.9
1963-64	110.2	105.9	108.4	97.7
1964-65	122.3	109.4	130.9	83.6
1965-66	131.6	117.0	141.7	82.6
1966-67	149.9	125.2	166.6	75.2
1967-68	167.3	129.1	188.2	68.6
1968-69	165.4	132.8	179.4	74.0
1969-70	171.6	139.7	194.8	71.7
1970-71	181.1	149.7	201.4	74.3
Base Year				
Weights	*(100.0)*	*(49.87)*	*(40.4)*	–
1971-72	105.6	109.5	100.4	109.1
1972-73	116.2	121.9	110.3	110.5
1973-74	139.7	139.5	139.2	100.2

Contd...

Year	General Index of Wholesale Prices	Index of Manufactured Products	Index of Agricultural Commodities	Prices of Manufactures as Per Cent of Agricultural Commodities
Base Year Weights	(100.0)	(49.87)	(40.4)	--
1974-75	174.9	168.8	165.9	99.3
1975-76	173.0	171.2	157.3	108.8
1976-77	176.4	175.1	158.4	110.5
1977-78	185.8	179.2	174.8	102.5
1978-79	185.8	179.5	171.9	104.4
1979-80	217.6	215.8	188.7	114.4
1980-81	257.3	257.3	210.8	122.2
1981-82	181.3	270.6	236.5	114.4
1982-83	288.7	271.1	248.3	109.6
1983-84	316.1	295.8	283.1	104.5
1984-85	338.4	319.5	303.2	105.4
1985-86	357.8	342.6	309.6	110.7
1986-87	376.8	359.4	330.1	108.9

The above table gives data on price movements during the three periods, viz. 1950-51 to 1961-62 to 1962-63 1970-71 and 1971-72 to 1983-84. From the price trends prevalent in the first period, no concrete inference can be drawn since the prices of manufactures as per cent of the prices of agricultural commodities were more than-100 in 4 years and less than 100 in 4 years.

However, in the second period (1962-63 to 1970-71), this percentage has been less than 100 for all years which clearly points to the fact that agricultural terms of trade have been favourable during this decade. In recent years, i.e., from 1971-72 to 1986-87, the prices of manufactured goods seem to be rising faster than the prices of agricultural goods. Except for one year, the relative manufactured prices have been higher than the agricultural prices. The terms of trade, therefore, seem to have gone against the agricultural sector.

The above method of determining the terms of trade between agricultural and manufactured sectors, however, has been criticised by many economists. Kahlon and Tyagi note that "the index of manufacturers" and "index of agricultural commodities" are inclusive of many such commodities which are not transacted between the two sectors at all.

Moreover, according to them, "the weights assigned to different commodities in the construction of the wholesale price index are on the basis of the total volume of transactions and do not reflect the quantities sold by one sector to the other. Thus, they do not truly represent the composition of trade between agricultural and non-agricultural sectors which is so important for the proper determination of the terms of trade."

Kahlon and Tyagi do not agree with the results of many studies made on the terms of trade between the agricultural sector and manufacture sector because "most of the studies suffered from serious limitations on account of limited coverage, use of improper weights, inappropriate price indicators, adoption of incorrect method for estimating volume of exports and the use of a method for constructing price indices which on a prior reasoning would underestimate the rise in prices of non-agricultural goods and inflate the rise in prices of agricultural commodities".

Further, much depends on the selection of base and terminal years. For example, if a drought year during which agricultural prices generally rule high is taken as the base, the terms of trade in the following years would seem to be moving against agriculture. On the contrary, if the base year with low agricultural prices is chosen, the terms of trade would move in favour of agriculture in the following years.

Instead of using general price indices, Kahlon and Tyagi in their study have used 32 items representing the purchases by the agricultural sector from the manufactured sector and 22 commodities to represent the basket of goods sold to the agricultural sector. Instead of using the index number of wholesale prices, these authors have used the farm harvest prices as indicators of prices received in the case of cereals, oilseeds, gur and tobacco. Kahlon and Tyagi have presented three sets of tables in their study based on Laspeyre's approach, CSO approach and Paasche's approach respectively.

The indices of prices received and prices paid along with the terms of trade calculated following the methodology evolved by Tyagi and Kahlon in 1980. It would be seen from the table that during the period 1952-53 to 1963-64, the terms of trade remained against the agricultural sector. During this period, the terms of trade have fluctuated between 73 to 91 and for the period as a whole, averaged at 85.

During the period 1964-65 to 1974-75, the terms of trade have remained in favour of the agricultural sector in almost all the years. It would be noticed that during this period, the rise in the prices of commodities sold by the agricultural sector for final consumption has been faster than the rise in the prices of commodities sold for intermediate consumption.

The terms of trade, however, again moved against the agricultural sector from 1975-76 onwards. It would be seen that the terms of trade remained between 83 and 91, i.e., though these remained against the agricultural sector but in most of the years, these were less adverse than in the earlier period.

In order to comprehend better, one has to know about public sector outlays and expenditure in details. The following tables show the same.

Public Sector Outlays and Expenditure during Ninth five year Plan (1997-98 to 2001-02), Tenth Plan (2002-03 to 2006-07) and Eleven five Year Plan (2007-2012) under various Heads of Development

(Rs. Crore)

Sl. No.	Sector	Ninth Plan Outlays$	1997-98 BE	AE**	1998-99 BE	AE**	1999-2000 BE	AE	2000-2001 BE	AE**	2001-02 BE	AE**	10th plan Outlays	2002-03 BE	AE**
1	2	3	4	5	6	7	8	9	10	11	12	13	14	15	16
1	Agriculture and Allied Activities #	42462	6974	5929	8687	7698	8796	7365	8281	7577	9097	8248	58933	9977	7655
2	Rural Development	74686	11156	10074	12203	10985	12252	11281	11349	9852	13444	14235	121928	15778	19753
3	Special Area Programmes	3649	781	874	1535	1184	1630	1514	1022	1045	1146	919	20879	1046	1066
4	Irrigation and Flood Control	55420	11220	9905	13444	10814	15672	14210	16234	13529	16528	14552	103315	16964	11965
5	Energy	222375	37111	31793	45134	35572	43273	35810	44856	36613	49103	37145	403927	53781	44710
6	Industry and Minerals	65148	13319	10306	13901	7979	10918	7248	11250	6866	10487	7942	58939	10608	8776
7	Transport	119373	20661	18101	25423	20347	27104	23463	31271	25734	35834	29918	225977	44258	35244
8	Communications	47280	13367	10131	14887	11376	16822	14039	19473	14183	20300	18083	98968	19551	13057
9	Science, Technology and Environment	18458	2431	2004	3289	2443	3471	2942	3927	3557	4042	3670	30424	4783	4160
10	General Economic Services	14580	2403	1811	3588	3071	4601	2452	4607	2538	8929	2948	38630	4557	4995
11	Social Services	183273	34260	26867	41359	38738	45154	38439	48368	41336	56373	46474	347391	63006	56954
12	General Services	12496	2221	1961	2458	1374	2569	1847	2721	1649	3608	2181	16328	3588	1868
	Total	**859200**	**155905**	**129757**	**185907**	**151581**	**192263**	**160608**	**203359**	**164479**	**228893**	**186315**	**1525639**	**247897**	**210203**

Sl. No.	Sector	2003-04 BE	2003-04 AE**	2004-05 BE	2004-05 AE**	2005-06 BE	2005-06 RE***	2006-07 BE	2006-07 RE**	11th Plan Qutlays	2007-08(#) BE	2007-08(#) RE**	2008-2009 BE
1	2	17	18	19	20	21	22	23	24	25	26	27	28
1	Agriculture and Allied Activities	9948	8818	11109	10778	13840	13440	16163	16987	136381	8558	8544	10075
2	Rural Development	16029	19858	17634	18579	23931	27214	30711	30458	301069	16706	17511	18972
3	Special Area Programmes	1848	1733	4939	2381	4547	4904	5522	5840	26329			
4	Irrigation and Flood Control	17751	14499	21111	19471	26957	25007	33189	32120	210326	507	411	411
5	Energy	59689	60174	65965	60375	77356	71306	90499	88296	854123	79158	93815	93815
6	Industry and Minerals	10101	7758	10756	10121	15303	13512	18213	17017	153600	20434	28836	28836
7	Transport	42309	36186	44481	38690	59992	58915	72055	74625	572443	71589	68930	84177
8	Communications	14914	12897	11764	9279	12226	17773	20366	18121	95380	25812	16599	21937
9	Science, Technology and Environment	5047	4417	6030	5513	7520	6411	8395	7258	87933	8816	7742	9283
10	General Economic Services	7123	4502	13462	5067	10080	7146	10026	8691	62523	3632	3043	6052
11	Social Services	68946	62621	77549	79244	106370	103308	131989	127738	1102327	83951	78798	100778
12	General Services	2179	3201	2270	3266	3119	2695	4159	4087	42283	829	533	1149
Total		**255883**	**236664**	**287070**	**263434**	**361239**	**351629**	**441285**	**431238**	**3644718**	**319992**	**324762**	**375485**

$: Indicates Ninth plan realaization.

** : For the Centre some of the revised estimate figures have been taken in to account for Actual Expenditure figures as the figures for some of the sub-heads are not available.

*** : Revised Expenditure has been taken into account for some Head of Devilopment for Centre and State in place of actual Expenditure.

: For Centre only,as figures for States & Uts are not available yet. Only Revised figures for centre has been given in place of Actual Expenditure.

: For Centre only,as figures for States & Uts are not available yet.

Source: Planning Commission, New Delhi.

Share of Public Sector Outlays and Expenditure under Agriculture and Allied Activities during Ninth Plan (1997-98 to 2001-02), Tenth Plan (2002-03 to 2006-07) and Annual Plan (2007-08)and Eleventh Plan (2007-12)^

(Rs. Crore)

Five Year Plan/Annual Plan	Agriculture and Allied Activities*		Total		Percent Share of Agriculture and Allied Activities to Total	
	Plan Outlays	Actual Expenditure	Plan Outlays	Actual Expenditure	Plan Outlays	Actual Expenditure
1	2	3	4	5	6	7
Ninth Plan (1997-02)	42462	37239$	859200	941041$	4.9	4.0
Annual Plan (1997-98)	6974	5929	155905	129757	4.5	4.6
Annual Plan (1998-99)	8687	7698	185907	151581	4.7	5.1
Annual Plan (1999-00) **	8796	7365	192263	160608	4.6	4.6
Annual Plan (2000-01) **	8281	7577	203359	164479	4.1	4.6
Annual Plan (2001-02) **	9097	8248	228893	186315	4.0	4.4
Tenth Plan (2002-07)	58933	56997	1525639	1489395	3.9	3.8
Annual Plan (2002-03) **	9977	7655	247897	210203	4.0	3.8
Annual Plan (2003-04) **	9948	8818	255883	236664	3.9	3.7
Annual Plan (2004-05)**	11109	10778	287843	263434	3.9	4.1
Annual Plan (2005-06)**	13840	12760	361239	347857	3.8	3.7
Annual Plan (2006-07)***	16163	16987	441285	431237	3.7	3.9

Contd...

Five Year Plan/Annual Plan	Agriculture and Allied Activities*		Total		Percent Share of Agriculture and Allied Activities to Totcl	
	Plan Outlays	Actual Expenditure	Plan Outlays	Actual Expenditure	Plan Outlays	Actual Expenditure
1	2	3	4	5	6	7
Eleventh Plan (2007-12)	136381	NA	3644718	NA	3.7	NA
Annual Plan (2007-08)#	8558	8544	319992	292337	2.7	2.9
Annual Plan (2008-09)##	10075	NA	375485	NA	2.7	NA

* Includes Crop Husbandry, Animal Husbandry, Dairy Development, Agricultural Research & Education, Forestry & Wild life, Plantation, Agicultural Marketing, Food Storage & Warehousing,Soil and Water Conservation, Agricutlural Financial Institutions, other Agricultural Programmes and Cooperation.

** For the Centre, some of the revised estimate figures have been taken into account for actual expenditure figures, as the figures for some of the sub-heads are not available.

*** - Revised Expenditure has been taken into account for some Head of Development for Centre in place of actual Expenditure.

For Centre only, as figures for States and UTs are not available yet. Only Revised figures for Centre has been givien.

For Centre only ,as figures for States and Uts are not available yet.

^- Data pertain to eleventh Five Year Plan are as approved by the National Development Council (NDC) in its meeting held on 19th December,2007.

NA : Not Available

Source: Planning Commission, New Delhi.

The movement of net barter terms of trade have a significant impact on the pace of investment in the agricultural sector as well as on the rate of growth of agricultural output. "The build-up of production potential stands adversely affected as a consequence of terms of trade turning against the agricultural sector, when it is not possible to shift the production function upwards due to non-availability of new technology.

The changes in terms of trade can also have a perceptible impact on the demand of non-agricultural commodities by the agricultural commodities by the agricultural sector. However, the impact of terms of trade turning adverse to the agricultural sector on the growth of investment in agriculture and on its output can be neutralised through technological developments".

9

Agricultural Labour

Labour in agriculture constitutes one of the most important factors of production as machine and labour cannot be perfect substitutes in the agricultural production process. In spite of the significant technological advance in this sector, man will continue to be more important here than in the rest of the economy. The association of labour with agriculture is as old as the farm occupation itself. A discussion on labour in agriculture shall have to be developed in a two-dimensional format. On the one hand, man (labour) is a consumer and on the other, he is a producer. This imbalance between how much he produces and how much he consumes has always placed countries differentially. If P denotes the amount produced and C, the amount consumed, then the relationship between P and C can be expressed as:

$$D = P - C = O$$

A positive D is an indicator of a farm sector producing more than what is consumed by the farm labour force and hence the surplus can be used in the non-farm sector. Zero or negative values of D are not in tune with the process of economic development.

The supply of labour and the demand for it form yet another important line of discussion on the subject. The inequality between the supply of and demand for farm labour has led to the emergency of several development issues at different stages of economic transition. Similarly, the quality of farm labour and its characteristics play a crucial role in the agricultural sector. Sensitivity with respect to seasonal variations makes the problem of agricultural employment a complex one.

Role of Labour in Agriculture

Labour is the single most factor which is of primary importance in increasing the production in traditional agriculture. The oldest production industry known to mankind is agriculture and the basic input in this production process has been human labour. At the early stage of human development, since land was abundant, increase in farm labour led to the clearing of more land and bringing it under cultivation. Hence, the volume of agricultural production was directly related to the volume of human labour and multiplication of labour units augmented the returns from this industry. Such a situation must have continued even after small urban settlements developed because farm labour in that case had not only to produce for those directly dependent upon agriculture but even for urban dwellers.

Transformation from traditional static agriculture to dynamic agriculture depended as much upon the farm labour force as on technological innovations. Many a technological innovation have been such, as would require added labour. Even with the use of increased capital, a few agricultural operations require intensification of labour such as seedbed preparation, weeding and harvesting, etc. Labour, therefore, is of critical importance on agriculture whatever the stage of its development may be.

Farm labour has been a pivotal point the model of economic development, in general, and in the models of agricultural development, in particular. The history of economic transition suggests that over a long period of time, the pressure of population on land increased considerably. Land being limited, with every increase in population, the man-land ratio went on increasing and consequently remuneration per unit of labour declined accordingly till it became constant at the subsistence level since it could not go below that.

In the absence of alternative employment due to non-existence or relatively small size of non-farm sector, agriculture had to absorb the shocks of additional manpower, rendering, in the process, a sizeable amount of farm labour as redundant. Development economists have done elaborate exercises to explain how this redundant labour force can be used to expand the capitalist sector and hence promote economic development of a country.

One of the significant models of economic development of a surplus labour economy is that of Lewis. It builds the development process on shifting the constant wage-earners from the agricultural sector to the non-agricultural sector at the subsistence wage and allowing the capitalists to appropriate huge profits for further investments and hence for the expansion of the capitalists sector. In this, Lewis-type models visualised double benefit from agriculture. First it serves as a supplier of cheap labour and secondly, as a supplier of food to the non-agricultural sector which in the process transfers income from the urban to the rural sector where after meeting the immediate consumption requirements, it goes into agricultural capital formation. Thus, labour plays a significant role in agriculture directly and indirectly as well.

Historical analysis of economic growth marks three distinct phases in the process of economic transformation. The first phase is when the survival of the economy is solely dependent upon agricultural employment. The sole contributor to national income is labour used in the primary sector whose productivity is undoubtedly low but has no other avenues of occupation on which to depend upon. The second phase begins with the improvement in farm productivity and development of a non-farm sector that draws labour from farms. Initially, non-farm jobs are mainly devoted to the production of materials used on the farm. With the introduction and expansion of non-farm jobs, the conditions of living improve resulting in the increases in size and quality of labour force. This further stimulates the growth of non-farm sector and as a result, continuously drags out labour from the farm sector. This then leads to the third phase of development marked by a rapid rise in non-farm employment and a higher rate of economic growth.

It is, therefore, easy to understand the significance of the relationship between labour and agriculture in its historical perspective. Directly or indirectly, it has played a crucial role in this primary sector of production.

Supply of Labour

Supply of labour force in agriculture depends upon the size of population. In traditional static agriculture, the size of farm labour force directly varied with the size of total population. Alternative employment avenues being non-existent, an increase in population straightaway led to an increase in labour force in the farm sector and *vice-versa.*

In the event of farm and non-farm sectors existing side by side, the agricultural labour force depends upon the residual formed by the rate of growth of population and the rate of growth of non-agricultural employment. The state of economic transformation from traditional static to modern dynamic society greatly influenced the supply of farm labour force.

On the demographic side, we shall have to examine different theories of population growth and the factors affecting it.

Theories of Population Growth

Theories of population growth are based on two different schools of thought – naturalistic school and sociological school. The former emphasises factors like fecundity, fertility, sterility, supply of food and physical resources, density of population and the standard of living as influencing the growth of population. Latter school of thought lays stress on institutional setup devised by human beings and emphasises social rather than environmental factors as influencing population growth. Cultural pattern and social behaviour are considered more dominating than the natural factors.

The thought developed by the naturalistic school is founded on Malthusian theory of population growth. This theory is based on the belief that the size of population can be limited by man himself while the capacity to produce food is limited by nature. His main contention was that the rate of growth of population would always outpace the rate of growth of foodstuffs and in order to maintain balance between the size of population and the volume of food availability, nature, will come down on population with a very heavy hand in the form of famines, pestilence, epidemics, etc.

Malthus, in his thinking obviously faltered on two points, viz., on man's inventive competence and on his acceptability of a cultural pattern of a controlled birth rate. Malthus did not foresee the possibilities of a technological revolution enabling man to produce more with less land at his disposal.

On the other hand, the social school of thought believed that man, while taking decisions regarding the size of the family, 'shall carefully consider the place of children in family-life goals and shall, in the process, be affected by institutional setup such as religion, social ceremonies and law, the nature and place of work of the women and parental motivation and expectations from their children. Besides, the size of population can also be regulated by formulating public goals relative to population. These goals may be so fixed as to slow down the rate of growth of population, to prevent or encourage an increase in population, to improve the quality of population and to secure a distribution of population relative to natural resources in different regions.

On the side of population growth, therefore, there is no limit to the supply of labour force to agriculture. In fact, the world is seized with the problem of over population relative to its requirement and in spite of the voluntary measures to regulate the growth of population, there appears to be no danger of short supply of labour force in the near future. Alarmed by the gravity of current trends in population growth, the suggestions for zero rate of growth are already put forth. The suitability or otherwise of such a strategy is being debated and investigated.

Factors, Affecting Population Growth

The size of population at a point of time depends upon the difference between births and deaths and corrected by the difference of in-migration and out-migration from that geographical territory. For the world, as a whole, the migrations are of no effect. Factors affecting population growth are, therefore, those that affect the births and deaths. Allowing for differential applicability and acceptability owing to differences in sociocultural background, religion and geography, the main factors that effect births are as follows:

1. Marriages.
2. Reproductive capacity.
3. Economic conditions.
4. Social, cultural and religious influences.
5. Level of education.
6. Government policies.

Marriages play a significant role as a factor affecting population growth. Three points have to be noted in this regard; number of marriages, the age at the time of marriage, particularly of females and the divorce rate and incidence of legal separations.

Populations with legal, religious and social sanctions for remarriages and more than one marriage with previous spouse or spouses living are prone to grow at a faster rate than other populations where these sanctions are not in vogue. Similarly, marriages at younger ages enlarge the reproductive age span and hence a higher rate of growth of population. Easy divorces and legal separations act as depressors on population growth and the extent to which the rate of growth can be limited is directly related to the intensity of these occurrences.

Reproductive capacity, that is, physical capability of couples to produce children (also known as fecundity) is an equally important factor affecting the number of births. A higher reproductive capacity with a larger reproductive span shall obviously lead to higher rates of population growth and *vice-versa.* Economic conditions of people are said to influence their decisions with regard to births of children. It is generally believed that better economic conditions are followed by lower birth rates.

The reason being that people do not want to allow their standard of living to fall owing to the larger family size and that with better economic conditions, more time is devoted to recreation. Social religious and cultural influences also have a considerable effect on the birth rate. Societies which consider birth control and abortions, etc. anti-social and anti-religious record a higher birth rate compared to other societies. Education too plays a significant role. Lastly, government can always formulate policies to regulate the birth rate. It the government is interested in checking the birth rate, it may withdraw incentives like free education to the third child, subsidised rations, free medical facilities, etc. Measures contrary to these may be taken in case government wishes to increase the birth rate.

Factors, Affecting Death Rate

1. Level of medical technology.
2. Economic conditions.
3. Infant mortality rate and expectation of life.

The level of medical technology is perhaps the single most factor influencing the death rate. Acceptability of advanced medical technology by the people is very crucial. Death rates have been reduced considerably by advances in medical research. Better economic conditions help reduce tensions on human mind and provide improved living conditions and socio-economic atmosphere. This goes a long way in reducing the death rate. Lower mortality rate among infants and older people enlarges the reproductive capacity of a society and may lead population to increase.

On the demographic side, therefore, the supply of farm labour force is directly linked with the overall size of population which, as we have seen, depends on the size of birth rates and death rates and, of course, on in-and-out migrations from a locality, country or region.

Considering the supply of farm labour force as a function of residual of population growth on the one side and growth of non-farm employment, on the other, it shall depend upon the nature and rate of expansion of the non-agricultural sector. Further, the growth of non-farm sector is likely to generate more employment and with stagnant rate of population growth, the residual shall tend to decline and hence restrict the supply of farm labour force. But note shall have to be taken of the nature of growth in the non-farm sector. If it uses capital-deepening devices, the rate of growth of employment may fall far short of the rate of growth of population and hence the residual will tend to increase and in the process, increase in the supply of farm labour force.

The developing countries of the world that have chosen to expand the non-farm sector at a faster rate are exactly faced with such a problem because the non-farm sector under the compulsions of market and international standards has used more capital deepening devices and failed to generate employment at a higher rate. In any case, even when non-farm employment expands at a rate equal to the rate of population growth, the technological advances would enable the production of agricultural products at the desired levels with lesser manpower.

Demand for Labour Force

Demand for agricultural labour force is influenced by the size of the farm, the system of farming and the degree of technological development.

In general, it would appear that larger the size of farm, the greater will be the requirement of labour and *vice-versa.* But in practice, it does not work in that fashion always. To some extent the direct relationship between farm size and labour requirement works all right but beyond that it may turn out to be otherwise; the turning point shall, of course, demand upon the system of farming and technological development. Small farms normally require more labour for a holding of a given size than large farms practising the same type of farming.

The system of farming is equally important. Intensive system of farming would, in general, require more labour per unit of land than is required under an extensive system of farming. In the light of our earlier argument, it follows that intensively-operated small farms would require a larger amount of labour force. Farming is such an enterprise where the required standards cannot be fixed. Besides size and system of farming, there are a variety of other factors that cause variations in the labour requirements. For example, the cropping pattern, dairy and poultry farms and horticulture, etc. have different labour requirements.

An acre of rice cultivation may have a different requirement of labour compared to an acre of wheat cultivation or potato cultivation and so on. Labour requirement on vegetable farms may significantly differ from that on floriculture and horticulture. Secondly, sensitivity of individual crops suddenly gives rise to more or less labour requirements during particular operations. Labour requirements for keeping livestock differ with the type of livestock kept and the system of management. There will be lesser demand for labour if cows are to be looked after in an outdoor pasture area than these are to be looked after indoors. Mechanical system of milking cows may generate lesser demand for labour than indoor hand milking system.

The nature and operational character of farming is such that precise labour requirements cannot be estimated. The services of a worker or a group of workers of different specialisations may be needed every day though not for a full day on small farms while on large farms, the situation may be different. Seasonality of agricultural employment is a very serious problem. This subjects the demand for labour to variations from one operation to another and from one seasonally more sensitive crop to another of lesser seasonal sensitivity. Shortage of labour is felt in peak seasons and labour is regarded surplus in slack seasons.

In regions where farming is predominantly dependent upon the monsoons, variations in demand for labour occur due to the nature and amount of rain recorded. If rains come on time, the agricultural operations run as scheduled and accordingly generate demand for labour on the usual pattern. Untimely rains cause variations in labour requirements as well. If, on the other hand, rains fail in a particular season, the demand for labour is completely upset and farm labour suddenly becomes redundant. On the contrary, acute shortage of labour is felt during the crop operations closely linked with rainy season.

To avoid shortages of labour and labour redundancy, farming has to be managed more scientifically and in a way so as to permit reasonable adjustments as and when required to avoid waste and overburden. Such a strategy can be worked out in two ways. The farm enterprises may be selected in such a combination so that the regular workforce finds full-time job every day and every season. Alternatively, a minimum regular workforce may be maintained and supplemented by additional stock as and when required. The two approaches involve the questions of farm management and labour management.

Traditional agriculture did not pay much attention to management problems as such but relied heavily on the use of casual labour during peak seasons. With advanced levels of economic development, the casual labour disappeared owing to expansion in non-farm employment, as is the case in the developed countries at present. This stimulated the need for better farm and labour management. In the developing countries of Asia, however, casual labour is available and used during the peak seasons. But the overall considerations of farming enterprise demand that farm management be strengthened and improved for obtaining better results from agriculture.

In many cases, peak agricultural operations are managed by pooling of labour force, particularly on small farms. This practice is in vogue in many Asian villages and shall continue to be there for quite sometime. Besides, labour problems are also solved by using contract labour and overtime labour. There is an advantage in using contract labour as it is self-organised and does not need permanent supervision as in the case of casual labour.

Overtime labour has a special place in agriculture owing to its peculiar nature compared to industrial production. Some of the farm operations have to be completed within a specified time dictated by nature and in such cases, longer hours of work have to be put in if casual or contract labour is not available. Cultivation of land, sowing, weeding and harvesting are such operations as would necessitate the use of overtime labour since the operations have to be completed in time. Similarly, keeping and supervising the livestock needs overtime labour because several jobs have to be done at fixed times and in a regular programme. There is, however, a limit to the use of overtime labour imposed by the physical capacity of a worker.

In these days, however, emphasis is being laid on farm management to ensure optimum-utilisation of farm labour force. Cropping pattern is so designed as to create demand for labour in one crop when it is relieved from the cultivation of another crop. Even daily adjustments are possible by combining small farms of vegetable cultivation poultry and dairy units with major crop operations. In the case of seasonality, non-farm employment is created by setting up small agro-based industrial units in the village itself.

We have explained how the size of farm and system of farming influence the demand for labour in agriculture. We shall now turn to the third factor, viz., level of technological development. The nature of technological progress greatly influences the demand for farm labour force. The history of the developed countries reveals that technological progress in agriculture tended to reduce the demand for farm labour.

The farming operations became more mechanised and displaced labour for use in non-farm sector, perhaps the Western industrialised countries and the United States were under compulsion to do so as the rates of population growth were low compared to the rates of growth of employment. Japanese experience shows that technological advancement in agriculture can be labour-intensive in character. It depends upon what input-mix and what type of machines are used on farms. Japanese developed an intermediate technology where inputs like chemical fertilizers and mini-tractors stimulated greater demand for labour rather than displacing it. New agricultural technologies result in income effect in that the efficiency of labour is increased and real income gains increase. This income effect is associated with a substitution effect.

The temptation to maximise gains from new technologies, and as a result to enjoy more leisure leads to the substitution of machine for man and displaces labour from the farm. On the other hand, if it suits to avoid heavy cost of machines and their maintenance, an increased dose of labour may be put in to optimise returns from new technology through intensified agricultural operations.

There is a strong controversy going on whether new agricultural technologies have displaced farm labour or intensified its use. Ample evidences for and against this argument have been given but no definite conclusions can be drawn. The nature and level of technological progress shall, however, continue to considerably influence the demand for farm labour force.

Efficiency of Farm Labour

The simplest measure of labour efficiency is the amount of output per worker. Productivity of farm labour is directly linked with the size of labour force, the quality of labour force and the nature and level of technological development. Changes in the productivity of labour are connected with the process of transformation from static to dynamic agriculture.

Size of Labour Force

In a static agricultural technology, as the size of labour force increases, the marginal productivity of labour (or more appropriately of additional labour) tends to decline. This happens because land productivity does not increase owing to the static technology and farming is extensive rather than intensive in character. If technology continues to be static and labour force rises continuously, the productivity of labour is compelled to fall with the result that wages fall below minimum subsistence level and as a consequence increase the death rates and reduce rate of growth of population.

Given the nature of agricultural production function, an initial dose of additional labour input may be rewarded by more than a proportionate rise in agricultural production and average produce of labour may tend to be higher than the subsistence level. Such a situation, for sometime, provides for the maintenance of even those members of labour force whose marginal product is below subsistence level. This form of farm labour productivity is typical of many developing countries where the levels of subsistence living are quite low. Such a phenomenon may be explained by the following figure.

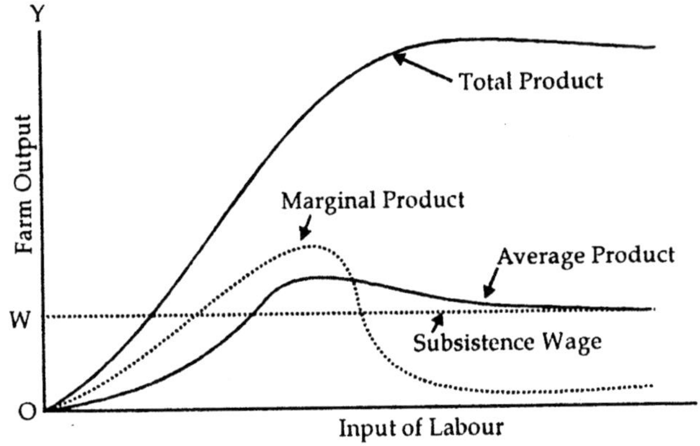

Fig.: Labour Productivity

It is evident from above figure that the level of subsistence living *OW* is relatively low and average productivity of labour is much above this level. This is particularly relevant if land productivity is high on new lands brought under cultivation by the additional labour force. But if the soil quality is poor and the possibility of extensive farming is limited, under static technology, the rise in labour force will not permit the average labour productivity to rise above the subsistence level. Such situations may often be found in the hill areas.

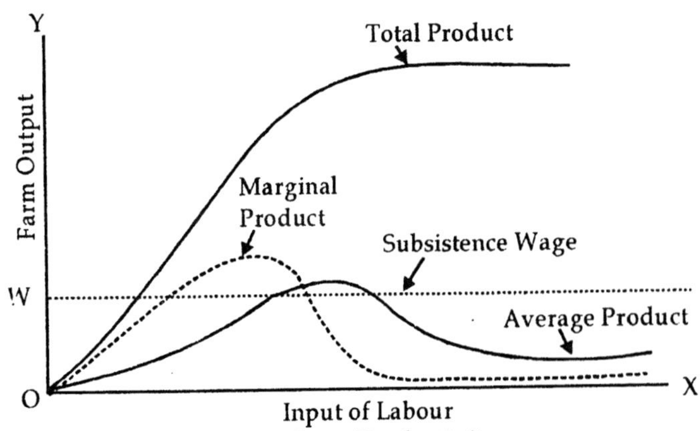

Fig.: Labour Productivity

In above figure, the subsistence requirement *OW* is as usual low. The average productivity curve does not rise beyond *OW* and increases in total product are also moderate.

As agriculture sheds its static character and tends to be dynamic, changes in technology may result in the rise of the productivity of labour even when the size of labour force is increasing.

Quality of Labour Force

Productivity of farm labour shall greatly depend upon its quality, both physical and mental. Physical quality of labour force refers to sound health, stamina and race characteristics. Differences in the health of labour force result in productivity differentials. Usually people living in temperate climatic conditions enjoy better health than those living in hot and humid climatic conditions. Sound health gives greater stamina to stand the odds of the occupation and to work overtime. Race characteristics are also a factor that influences quality of labour. Punjabis, for example, have shown exemplary physical fitness in bringing about a revolution in farming both in Pakistan and India. On the other hand, farmers living in the rest of the country and enjoying same farm conditions have not been able to keep pace with their Punjabi counterparts.

Mental quality of labour force shall depend upon the level of education, extension services and agricultural training programmes. Education, more particularly, functional and vocational, rends to improve the management capabilities of farm labour and with the same amount of work they can achieve better rewards.

Extension services are equally important to narrow the gap between the farm labour and farm scientist. Unless laboratory results are made to reach the farmer, his mental horizon cannot be widened and his capability of enhancing farm productivity cannot be explored. Agricultural training programmes have a significant effect on improving the mental health of farm labour force. Agricultural universities and other such institutions are rendering useful service in this regard.

Level of Technology

The nature and level of technological development has a great influence on labour productivity in farming. Improved methods of farming, better quality seed, fertilizers and pesticides and improved tools and implements have resulted in considerable increase in farm labour productivity. As already mentioned, Japanese methods of cultivation are not only labour productivity enhancing but are also labour-intensive in character.

Such technologies increase employment per unit of land and also increase productivity per unit of employment. Replacement of outdated tools and implements have enabled better exploitation of soil qualities resulting in increased output per unit of labour. Replacing a plough with a tractor, in particular, and hand operations with mechanical operations, in general, have enabled the developed countries to raise several times the farm output with fewer farm workers.

Therefore, whether it is a labour-using and capital-saving technology, as in the Japanese case, or a labour-saving and capital-using technology, as in the case of developed west European countries and America, labour productivity can be enhanced enormously with technological development. The amount of capital with which a farm worker has to work determines the level of his productivity on different farms and in different situations.

Labour efficiency also depends upon how farm enterprises are combined, the diversification of production and supervision. Situations often arise in farming when the total workforce cannot be used full time in-one single enterprise and combination of enterprises have thus to be chosen in such a way that the labour force is optimally used. The efficiency of labour will certainly be poor if such enterprise combinations cannot be worked out.

Modern agriculture has no longer remained static in character and possibilities of multiple cropping have increased owing to new seed-fertilizer technology. This has also enabled the optimal use of farm labour force and resulted in better labour efficiency. Diversification of crops to suit different seasons is another way of increasing labour efficiency by utilising the total workforce for full time through different seasons. The management and supervision of labour makes a given amount of labour more effective on one farm than on another. Factors like distance of the farm from the place of residence or the distance between the parcels of land also affect labour efficiency.

Agriculture differs from manufacturing industry in that it offers little possibilities of division of labour as is the case in industry. The nature of farm enterprise is such as would demand greater sense of responsibility and willingness from farm workers. Farm workers are also expected to exhibit a greater range of technical ability as a single worker may be expected to do some manual work, some mechanical work and some supervisory work as well. Skilled farm work, therefore, is not the same thing as specialised work in industry. Farmers' technical ability will depend upon his training and technological development.

Farm worker's efficiency is also affected by his family background. A worker belonging to a farm family will tend to be more efficient than the worker coming from a non-farm family as the former enjoys the benefit of his early experience and also the experience of his family. The ratio of farm family workers to hired workers coming from non-farming background greatly affects the efficiency of labour. Gains or losses in labour efficiency obviously depend upon whether the ratio of experienced workers to inexperienced ones is favourable or not. Besides, age distribution of the farm labour force also effects the labour efficiency.

Further, farm transport greatly increases the mobility of farm labour and makes the farm more accessible to market.

Crop management is greatly affected by the introduction of tractor and other mechanical equipment. Introduction of mechanical operations in livestock and dairying have facilitated the processes of milking, cooking and sterilising. Technical improvements have, therefore, greatly influenced labour efficiency by pervading in all fields of farming.

The changes due to technical improvements occur in two different forms. Firstly, for the same produce, fewer hands are required and lower costs may have to be incurred than before. This directly enhances the output per unit of labour. Secondly, with the same amount of labour, a greater absolute amount of output may be realised which also enhances the output per unit of labour. Mostly mechanical improvements take the shape of labour-saving devices and may also enhance the volume of output. To this extent, the double effect of mechanical improvements raises the labour efficiency significantly.

Lastly, it may be pointed out that mechanical improvements alone do not raise labour efficiency. The other supplementary advances in the productivity of crops and livestock are equally important in this regard. Improvements in crop productivity change the ratio of output to employment and hence result in the increased labour efficiency. The same can be said about the output of livestock. The modern methods of plant nutrition and pest control, on the one hand, and the provision of better feed and breed and also control of animal diseases, on the other, have increased crop and livestock productivity to the advantage of the labour force. The efficiency of labour has accordingly increased.

Agricultural Wages

The farm labour force consists of family farm workers and hired farm workers. While as family workers have not to be paid directly, wages have to be paid to hired workers. Under conditions of static technology in traditional agriculture, farm wages in the past were determined by the subsistence theory. This theory provided for the fixation of agricultural wages in such a way as would enable the worker and his family to live on a minimum subsistence level. Even legislative and administrative measures provided for the regulation of agricultural wages on the basis of this theory. This was also the basis of the Poor Law. At present, even though the norms of farm wage fixation have changed favourably for the farm workers, these continue to be less than the wages earned in the non-farm sector.

Historically, the gap between farm and non-farm wages have been justified on the ground that a rise in farm wages results in more than a proportionate rise in non-farm wages. Even the differences in farm and non-farm enterprises accounts for the difference in wages.

Adam Smith in his Wealth of Nations explained, at length, the reasons that account for wage disparities between farm and non-farm workers. Another feature of historical character is that agriculture draws its hired manpower from the residual mentioned earlier. Throughout history, it has been a universal feature of industrial growth to use capital deepening devices due to which the rate of growth of industrial employment has always remained far below the rate of growth of population resulting in the swelling of residual. This residual has willy nilly, to depend on agriculture for its survival and hence the wages in agriculture get depressed to the extent the volume of residual increases.

At the initial stages of industrial growth, the farm sector is the only supplier of urban workers and industry feels obliged to draw this labour force out by offering higher wages. This does not immediately raise the farm wages but as the surplus manpower is exhausted, the average productivity per farm worker rises and leads to a rise in wages. This process, however, does not continue for long because with the expanding industrial sector, technological developments in industry take place which, it has been observed, tend to displace labour from industry. This restricted scope of wage employment in the non-farm sector has always kept the agricultural wages low.

In the developing countries of Asia, the problem has become all the more serious owing to the demographic explosion. With improved sanitation, public hygiene and better nutrition, the rate of growth of population has tended to be far higher than the rate of growth of industrial employment. This has, therefore, lowered the average productivity of labour in farming resulting in their wages being lower than those prevailing elsewhere. The popular governments have, however, enacted legislations providing for fair wages to farm workers. Of their own, farm workers have poor bargaining power because of lack of organisation and trade unionism relative to industrial workers and cannot get justice without state intervention. State governments have therefore, been enacting legislations from time to time to save farm labour from exploitation. These legislations provide for fixing of minimum wages, defining employment for overtime payments and part payments to be made in kind. The wages are so fixed as would enable a farm worker to maintain himself and his family in reasonable comfort and to promote his efficiency.

State legislations in the developed countries of the world have improved the status and lot of farm workers considerably. The employer has to provide for the worker's health, provident fund, accident and insurance to ensure a better deal to the farm worker throughout. These have also promoted a uniform wage structure throughout the country. An encouraging impact of these legislations has been to make less efficient employers more efficient to enable them to play better wages to their workers.

In the developing countries, the governments shall have to provide for these measures as listed above to improve the lot of farm workers.

10

Agricultural Development and Foreign Trade

Indian agriculture even in its traditional form has contributed to foreign trade. Over the time, Indian agricultural products have been facing stiff competition from other Asian countries. In the wake of globalisation and liberalised regime, this competition is likely to increase further and new initiatives in agricultural development shall have to be taken to meet the emerging challenges. The performance of agriculture after integration with the world markets greatly linked to the success of exports. In its bid to increase overall exports, the Government of India has decided to achieve this objective by giving a push to production and export of agricultural commodities. Agriculture has been a source of foreign exchange in the past.

However, most of the export earnings from agriculture come from the conventional items such as tea, coffee, spices, tobacco and cashewnuts. The current strategy for agricultural exports puts emphasis on diversification of agriculture including production of fruit, vegetable and flower crops and expanded production of aquaculture. The current emphasis now is on the production and growth of non-conventional items of exports in agriculture, namely, cereals, fruits and vegetables.

Achievement of the country on this front, however, depends upon the extent the agricultural goods enjoy comparative advantage and competitive position in the world market. Each commodity produced in agriculture should, first have a competitive position before it can be launched for exports. Recently, in this respect, a comprehensive study has been made by Gulati and others of the National Council of Applied Economic Research (NCAER) to determine competitiveness of agricultural commodities. In their paper, they have made an attempt to examine the available measures of competitiveness of agricultural commodities from the standpoint of exports and determine factors which may contribute to export competitiveness of agricultural commodities.

Export Competitiveness Measures

In determining the export competitiveness, they have taken Indian prices of agricultural commodities and compared them with international prices. They have discussed competitiveness in two alternative situations, namely, "importable hypothesis" and "exportable hypothesis." Under the importable hypothesis, the commodity in question is regarded as an import substitute, *i.e.*, the commodity under consideration is assumed to be imported and expected to compete with the domestically produced commodity in domestic market. Thus, this approach includes the international price of the commodity under consideration, international transportation cost between exporting country and India and post-clearance charges.

Further, if the commodity is required to be transported for sale in the regional markets, then the transportation cost within India is also included in the price of the commodity. Under the exportable hypothesis, the Indian Commodity under consideration is expected to compete in foreign country at a foreign port. The relevant reference price is determined after deducting the domestic and international transport costs, marketing cost and traders margin as also processing cost necessary to make commodity marketable.

The commodity is judged as exportable if the price of domestically produced commodity is less than the international price of the commodity, Mr. Gulati have evolved a concept, namely Nominal Protection Co-efficient (NPC) and this concept is the measuring rod of competitiveness. The NPC of a commodity has been defined symbolically as:

NRC = Pd/Pb

Where, NPC = National Protection Co-efficient

Pd = Domestic Price of Commodity

Pb = Border or reference price of a commodity after taking care of transportation and marketing expenses.

Under both hypothesis, a commodity is adjudged competitive if the NRC is less than unity. In other words, if domestic price of a commodity is less than the border price or reference price inclusive of importable and marketing costs, the commodity is judged as export competitive. The dominant factors are the domestic price and transportation costs.

The authors have also evolved a concept namely, Degree of Export Competitiveness, which is an extension of the concept of NPC. Symbolically, the Degree of competitiveness has been expressed as under:

$$\text{Degree of Export Competitiveness} = \frac{1\text{-NPC}}{\text{NPC}}$$

The higher the ratio, the greater will be the degree of competitiveness of the commodity under consideration.

With the help of above concepts, the authors have discussed export competitiveness in respect of major cereals fruits and vegetables produced in India. Among cereals rice has been found as highly export competitive. Wheat has also become export competitive while maize and sorghum are less export competitive. The degree of export competitiveness of rice worked out by the authors is 1.2 whereas that of wheat is 0.2. The degree of competitiveness of maize and sorghum is worked out to be -0.1 and -0.25 respectively.

Among fruits and vegetables, banana, lychee, tomato, grapes, sapota, onion, mango and potato have turned out to be positive export competitiveness products.

The above findings of the authors are of great value to policy planners and entrepreneurs in India. They provide a basic for judging the competitiveness to an individual entrepreneur if he decides to go in for an export venture in agricultural goods.

There is a host of non-price factors which contribute to competitiveness of agricultural commodities. These factors relate to consumers tastes and preferences, brand equity of products, changes in consumers' taste and preference, packaging, etc.

Current Status of India's Agricultural Foreign Trade

The Indian agricultural sector with the significant share in GDP plays a significant role in the employment generation specially in the rural sector. In the aftermath of agreement on agriculture under the aegis of WTO it has significant potential as a net foreign exchange earner.

The main approach of the Government policy has been to control trade in a manner to ensure adequate availability of essential food items to consumers at reasonable prices and to protect farmers from foreign competition. The objectives of export and import policy with respect to agricultural foreign trade is given as under:

Exports: To maximise agricultural exports in order to ensure remunerative prices to the farmers and boost foreign exchange earnings, keeping in view the prime consideration of sufficient availability of essential commodities to the domestic consumers at reasonable prices.

Imports: To regulate imports, keeping in view the domestic demand and supply situation, indigenous production, export potential and consideration of foreign exchange.

Specific policy in respect of principal agricultural commodities as contained in EXIM policy 1997-2002 is given below:

Rice: Export is allowed freely without any restrictions now.

Coarsegrains: Exports are allowed upto a quantitative ceiling of 50,000 Mts. during 1997-98.

Wheat and Wheat Products: Export of wheat banned now, but export of wheat products, whether in bulk or in consumer packs, be allowed within an over all quantitative ceiling of 0.5 million tonnes.

Pulses: Export is permitted against a licence subject to a quantitative ceiling but export in consumer packs be freely allowed.

Hybrid Jowar: The export of hybrid jowar is freely allowed.

Oilseeds: *(a)* The exports of HPs groundnut and sesame seeds are freely allowed for the year 1997-98 *(b)* The export of castor seeds is allowed freely.

Tobacco: Export is free without any restrictions for both *(a)* unmanufactured and *(b)* manufactured.

Spices: Export is free without any restrictions.

Cashew: Export is free without any restrictions.

Horticultural, Floricultural and Fresh, vegetable products Exports are free without any restrictions.

Export Performance

Increased export of agricultural commodities has been witnessed over the years. In the last five years, the value of agricultural exports has increased from Rs. 78,84,29,00 thousands in the year 1992-93 to Rs. 2,10,20,99,00 thousand in the year 1996-97. The commodities like wheat, rice, groundnut, oilmeals and raw cotton have registered significant increase in exports during the last five years. Table given below provides information about agricultural exports of principal commodities during 1992-93 to 1996-97.

The statistics pertaining to agricultural exports of India clearly shows that increased exports of agricultural commodities has been witnessed over the years. In the last five years, the value of agricultural exports has increased from Rs. 78,84,29,00 thousand in the year 1992-93 to Rs. 2,10,20,99,00 thousand in the 1996-97. The commodities like wheat, rice, groundnut, oilmeals and raw cotton where significant increase in exports (in quantity) has taken place during the last five years.

Year wise data on volume and value of imports of agricultural commodities are given in the following table. The figures do not reveal any discernible trend. However, imports of agricultural products, in general, have been going down except for wheat which has considerably increased in 1996-97 compared to previous years.

The increased imports of wheat is an indication of wheat production reaching the plateaus which situation will have to be changed. New technology breakthrough in wheat production is required to meet the growing needs of the Indian economy. On the other hand gains through exports of rice (Basmati), spices, oil meals, sugar and molasses and marine products will have to optimised in fact in the liberalised regime emphasis will have to be placed on exports because imports of agricultural products should normally not pose any alarming situation.

Needless to say that this is all budget oriented. Hence, one has to be aware of the budget estimates and all sorts of expenditures. The following tables show it in details.

Budget Estimate, Revised Estimate and Actual Expenditure of Department of Agriculture & Cooperation from 1991-92 to 2007-08

(Rs.in Crore)

Year	*BE*	*RE*	*AE*
1	2	3	4
1991-92	1041.35	1016.93	957.86
1992-93	1050.00	1273.16	1214.88
1993-94	1330.00	1320.05	1183.50
1994-95	1405.00	1458.84	1279.43
1995-96	1490.00	1325.39	1246.42
1996-97	1471.25	1377.91	1378.61
1997-98	1431.00	1266.28	1222.80
1998-99	1956.00	1378.41	1358.89
1999-2000	1956.00	1492.00	1471.90
2000-01	1965.00	1692.00	1666.04
2001-02	1985.00	1985.00	1792.92
2002-03	2167.00	1667.00	1676.78
2003-04	2167.00	2120.00	2050.73
2004-05	2650.00	2945.00	2657.42
2005-06	4179.32	3890.00	3798.69
2006-07	4800.00	4860.00	4860.00
2007-08	5560.00*	6927.94*	7049.36*

BE: Budget Estimate, RE : Revised Estimate, AE : Actual Expenditure

* including State Plan Scheme.

Source : Plan & Policy and Budget Divisions, Department of Agriculture & Cooperation.

Budget Estimate and Expenditure of Department of Agriculture and Cooperation

(Rs. Crore)

Period	*Centrally Sponsored Schemes*	*Central Sector Schemes*	*Externally Aided Projects*	*State Plan Schemes#*	*Total*
1	2	3	4	5	6
IX Plan Outlays	5533.65	3343.17	277.00		9153.82
1997-98					
BE	913.65	505.35	12.00		1431.00
Expenditure	804.53	399.93	18.34		1222.80
1998-99					
BE	1348.08	564.19	43.73		1956.00
Expenditure	957.67	372.85	28.37		1358.89
1999-00					
BE	1166.35	735.64	54.01		1956.00
Expenditure	876.91	559.05	35.94		1471.90
2000-01					
BE	1063.76	847.24	54.00		1965.00
Expenditure	583.59	1062.60	19.85		1666.04
2001-02					
BE	1215.40	731.99	37.61		1985.00
Expenditure	801.97	962.74	28.21		1792.92
IX Plan Expenditure	4024.67	3357.17	130.71		7512.55
X Plan Outlays Proposed	7236.73	5694.00	264.27		13195.00
2002-03					
BE	1341.53	760.47	65.00		2167.00
Expenditure	839.58	737.94	79.26		1676.78
2003-04					
BE	1005.72	1149.25	12.03		2167.00
Expenditure	961.41	1065.72	23.64		2050.73
2004-05					
BE	1072.69	1543.81	33.5		2650.00
Expenditure	1676.23	958.53	22.66		2657.42
2005-06					
BE	2785.62	1390.61	3.09		4179.32
Expenditure	2336.21	1460.87	1.61		3798.69
2006-07					
BE	2556.90	2243.10	0.00		4800.00
Expenditure*	3253.11	1606.89	0.00		4860.00
2007-08					
BE	4093.40	1426.6	0.00	40.00	5560.00
RE	3444.49	2443.45	0.00	1040.00	6927.94
Expenditure*	3724.19	2038.28	0.00	1286.89	7049.36

* Provisional. # The State Plan Schemes introduced from 2007-08 are (1) Rashtriya Krishi Vikas Yojana. (ii) Watershed Development in shifting Cultivation area in North Eastern states. *Source* : Plan & Policy and Budget Divisions, Department of Agriculture & Cooperation.

Key Economic Indicators of India

Indicator	Unit	2003-04	2004-05	2005-06	2006-07	2007-08
GDP at Factor Cost						
at Current Prices	Rs. Billion	25382	28777	32757	37901*	
at 1999-2000 prices	Rs. Billion	22228	23884	26128	28643*	
GDP in Agcultural & Allied Sectors						
at Current Prices	Rs. Billion	5323	5524	6158	6954*	
at 1999-2000 prices	Rs. Billion	4827	4824	5110	5302*	
Population (as on 1st October)	Million Number	1072	1089	1106	1122	1138
Per Capita National Income (Net National Product at Factor Cost)						
at 1999-2000 prices	Rs.	18317	19325	20358	22533*	
Foodgains Production	Million Tonnes	213.20	198.4	208.6	217.3	227.3#
Inflation Rate						
WPI based (1993-94=100)	%	5.5	6.5	4.4	5.4	4.7
						7.9
CPI-IW based (1982=100)	%	3.9	3.8	4.4	6.7	6.2
WPI for Agricultural commodities		182.9	186.7	190.7	204.1	218.1^^
WPI for Non-Agricultural commodities		174.0	187.5	196.9	206.8	21.7^^

Contd...

Indicator	*Unit*	*2003-04*	*2004-05*	*2005-06*	*2006-07*	*2007-08*
Foreign Trade						
Export	US$ billion	63.8	83.5	103.1	126.4	111.0
Import	US$ billion	78.2	111.5	149.2	185.7	168.8
Balance	US$ billion	-14.4	-28.0	-46.1	-59.3	-57.8
Total Agricultural Imports	Rs. Crore	21972.7	22811.8	21499.2	29637.9	29777
% to National Imports		6.1	4.6	3.3	3.5	3.1
Total Agricultural Exports	Rs. Crore	37266.5	41602.7	49216.0	62411.4	77769.7
% to National Exports		12.7	11.1	10.8	10.9	12.2
Total Long - Term Debt (End March)	US$ billion	107.21	115.25	118.56	143.39	159.73 q
Forex Reserves (End March)	US$ billion	107.4	135.60	145.1	191.9	281.2$
ExchangeRate (at end of year)	Rs/US $	45.95	44.93	44.27	45.25	40.41 ^^

*Quick Estimates. # Fourth Advance Estimates. ^^ Average April -December 2007. ^End November 2007. $ Upto February,2008

Source :(i) Central Statistical Organisation. (ii) Directorate of Economics & Statistics, Ministry of Agriculture. (iii) Economic Survey 2007-08. (iv) Office of the Economic Adviser, Ministry of Commerce & Industry.

Disaggregated Public Sector Outlays and Expenditure under Agriculture and Allied Programmes during Ninth Five year Plan (1997-98 to 2001-02),Tenth Plan (2002-03 to 2006-07) and Eleventh Plan (2007-12)

No.	Programme	1997-98		1998-99		1999-2000		2000-2001		2001-02		2002-03	
		BE	AE**	BE	AE**	BE	AE**	BE	AE**	BE	AE**	BE	AE**
1	2	3	4	5	6	7	8	9	10	11	12	13	14
1.	Crop Husbandry	2059	1694	2655	1845	2569	2175	2785	2503	3402	3146	3361	2597
2.	Horticulture	-	-	-	-	-	-	-	-	-	-	-	59
3.	Soil & Water Conservation	566	561	825	638	932	806	706	885	787	629	1315	544
4.	Animal Husbandry	451	340	495	326	525	389	461	425	500	488	506	371
5.	Dairy Development	141	70	133	65	145	61	97	85	101	89	87	105
6.	Fisheries	341	284	397	291	410	323	331	319	346	268	314	269
7.	Forestry & Wildlife	1334	1264	1832	2491	1966	1695	1741	1740	1920	1458	2032	1520
8.	Plantation	206	136	168	189	156	118	182	175	147	193	177	211
9.	Food Storage & Warehousing	122	117	108	29	136	107	170	26	145	191	186	218
10.	Agriculture, Research & Education	575	541	819	675	888	810	922	851	1000	968	1026	939
11.	Agricultural Financial Institutions	489	154	499	402	303	297	90	108	103	83	57	82
12.	Cooperation	573	616	635	660	632	536	689	381	492	184	214	443
13.	Other Agricultural Programmes	118	153	122	88	134	50	108	79	155	551	703	297
	Total Agriculture and Allied Programmes	**6974**	**5929**	**8687**	**7698**	**8796**	**7365**	**8281**	**7577**	**9097**	**8248**	**9978**	**7655**

No.	*Programme*	*2003-04*		*2004-05*		*2005-06*		*2006-07*		*2007-08(#)*		*2008-09(##)*
		BE	*AE***	*BE*	*AE***	*BE*	*AE***	*BE*	*RE***	*BE*	*RE***	*BE*
1	*2*	*15*	*16*	*17*	*18*	*19*	*20*	*21*	*22*	*23*	*24*	*25*
1	Crop Husbandry	3726	3256	4240	3932	5624	4989	6196	6216	4657	4922	5902
2	Horticulture	-	105	118	166	189	177	335	314	-		
3	Soil & Water Conservation	893	667	976	767	941	669	1072	979	21	21	11
4	Animal Husbandry	545	451	661	634	909	777	1071	1185	540	467	620
5	Dairy Development	93	69	116	110	154	181	187	216	80	101	89
6	Fisheries	297	254	375	350	473	446	660	576	284	171	291
7	Forestry & Wildlife	2070	1729	2048	1758	2296	2256	2769	2797	627	694	751
8	Plantation	245	177	301	276	345	293	378	383	274	261	313
9	Food Storage & Warehousing	186	186	126	138	233	231	332	272	333	269	225
10	Agriculture, Research & Education	1020	1029	1257	1166	1434	1379	1768	1911	1458	1291	1584
11	Agricultural Financial Institutions	50	60	82	33	66	50	124	109	-		
12	Cooperation	583	389	494	454	618	512	829	530	124	142	137
13	Other Agricultural Programmes	232	446	315	995	557	800	442	1499	160	205	152
	Total Agriculture and Allied Programmes	**9940**	**888**	**11109**	**10777**	**13840**	**12760**	**16163**	**16987**	**8558**	**8544**	**10075**

Note : (1) Totals may not tally due to rounding off of the figures. $: Indicates Ninth plan realaization. ** : For the Centre some of the revised estimate figures have been taken in to account for actual Expenditure figures as the figures for some of the sub-heads are not abailable. *** : Revised Expenditure has been taken into account for some Head of Devilopment for Centre and State in place of actual Expenditure. # : For Centre only,as figures for States & Uts are not available yet. Only Revised figures for centre has been given in place of Actual Expenditure. ## : For Centre only,as figures for States & Uts are not available yet. *Source:* Planning Commission, New Delhi.

Future Strategy

On the agricultural exports front, consistent and persistent efforts to enhance agricultural productivity, processing and preservation facilities for agri-products are needed.

Intensive research on post-harvesting grading, packaging and quality improvement of exporting material is required. This includes extending necessary assistance on matters such as communication, packaging, warehousing, transportation and speedy customs clearance. It is also equally important to ensure coordination with various organisation both within the country and abroad.

Government of India has already taken various measures to boost farm exports. These include cash assistance import replenishment, duty drawback, abolition of exercise duty and imposition of export duties where necessary. Attention has also been given to strict quality control and preshipment inspection systems.

Though the debate on whether Indian farmers are responsive to prices and other incentives is an unsettled one, we should pursue with the conviction that the agricultural sector is responsive to incentives and that incentives are needed for export promotion. The basic question however is that the incentives should reach the producers. In view of the unorganised nature of the agricultural sector, it is to ensure that the incentives such as price support, export subsidies, concessional finance, etc.

Indian Agriculture and WTO

GATT (General Agreement on Trade and Tariffs) was founded in 1947 with the objective of prescribing the rules for international trade. This international body had initially 23 members including India. Over the years the membership of GATT swelled and in 1994 it had reached 118 member countries. The overwhelming majority of this body, roughly two-thirds, comprises of developing countries. The members of the GATT account for 93 per cent of world trade at present.

Several left over countries which are not members of this body are keen to seek admission in it. Prominent among these countries is China which is making all out efforts to become the member of this body.

The headquarters of the GATT is situated in Geneva. Besides other activities, GATT is akin to an internationals Court. With the result it is empowered to resolve trade disputes creeping up among its members.

The detailed rules of the GATT weave around the following three basic principles:

1. World trade should be carried on non-discriminatory basis.
2. Domestic industry should only be protected by means of custom tariffs and not through other commercial measures.
3. The GATT serves as a framework within which negotiations can be held to reduce tariff and other trade barriers.

GATT provided for reduction in tariffs and trade restrictions in a phased manner over a period of time. The guiding principle of GATT has been, what is known as the MFN (most favoured nation) clause. This clause says that, "any advantage, favour, privilege or immunity granted by a contracting party to any product originating in or destined for any country shall be accorded immediately and unconditionally to the like product originating in or destined for the territories of all contracting parties". The purpose of this clause was to encourage multilateralism so that the world trade could expand in an orderly manner. GATT also provides for action by the contracting parties for settling disputes that might arise due to the application of the agreement.

The GATT has proposed multilateral trade regime with the suggestions that its benefits will be experienced in due course of time by all member countries. The member countries meet periodically to take stock of progress and problems of international trade and these talks are usually referred to as GATT Rounds of Talks. Since 1947, eight rounds of talks have taken place and the latest round was held in Uruguay and it came to be known as Uruguay round of talks. This round assumed considerable importance and was spread over seven long years (1986-93). It was after this round that WTO (World Trade Organisation) was born and the expression GATT was given up.

In the normal course, the Eighth round of talks should have been concluded in December 1990, but in the absence of a concensus, the same did not happen. The then Director General of GATT, Mr. Arthur Dunkel (now retired) was assigned the role of preparing a draft proposals as a compromise formula. The package of proposals presented by Mr. Dunkel came to be known as Dunkel Draft and was later signed by all members on December 5,1993.

The Dunkel Draft is now part of the new GATT accord, approved at Marrakesh or April 15,1994. This has paved for setting up of World Trade Organisation (WTO) with effect from January 1,1995. The significance and comprehensiveness of the 8th round lies in the sense that it covers a wide range of areas like agriculture, TRIPS, TRIMS, etc., A noteworthy condition in the Dunkel Draft is that each and every clause must be adopted without leaving any requiring that the draft be accepted in its totality or not at all.

GATT Recommendations Relating to Agriculture

For the first time, agriculture was included in GATT. Following are the recommendations of GATT with respect to the agricultural sector.

1. Non-tariff barriers like quotas imposed by nations on foreign agricultural products will be converted into tariffs, which will be reduced by 36 per cent in case of developed countries and 24 per cent in case of developing nations. This tariff cut will be implemented

over 6 years for the developed countries and 10 years for the developing nations.

2. By 1999, the term subsidies should not exceed 10 per cent of the value of the output of a product. Trade distorting support for farmers will be cut by 20 per cent over 6 years for developed nations and 13.3 per cent for the developing nations.
3. The value of export direct subsidies will have to be cut by 36 per cent and the volume by 21 per cent over the next 6 years by the developed nations.
4. Countries having a closed market for agricultural products have to import at least 3 per cent of their domestic consumption.

On the basis of these recommendations the following four areas have been identified which are causing concern to India regarding its agriculture sector.

1. The accord will force the Government to reduce subsidies given to Indian agriculture.
2. The Government will have to import three per cent of the domestic demand for agricultural products.
3. The accord will interfere with the Government's policies on public procurement operations and the public distribution system.
4. It will require the patenting of seeds, forcing farmers to buy them from multinational firms.

The supporters of GATT proposals, on the other hand, believe that these will bring some advantage to Indian agriculture in the manner as described below:

1. India's exports will increase as the international prices of agricultural products will go up. This will help our country to add more foreign exchange to its reserves and guarantee our peasantry a good price for their products. India should exploit this opportunity to increase its exports of rice, fruits, milk products, cotton,

floriculture, processed food, etc. and optimise to gains from the emerging international market.

2. Patenting of seeds will benefit India as it will bring latest developed and high productive hybrid seeds and enable the farmers to increase their productivity.

Objective Evaluation

The apprehensions that in response to GATT proposals, the Government will have to reduce the subsidies given to the farmers are not well placed. As per GATT recommendations, the farm subsidies granted by any Government to its farmers should not be more than 10 per cent of the value of output of a product. It is providing minimum support price to 20 agricultural products in each of which the subsidy is well below 10 per cent. Taking into account the current levels of subsidy offered, the fear that there will be a cut in subsidy is unfounded.

But on the other hand the agricultural exporting countries like France, Australia, etc., have been paying subsidy ranging from 35 per cent to 90 per cent. If such a high level of subsidy comes down, the prices of agricultural products in those countries will shoot up giving ample opportunity for Indian farmers to enlarge export and enjoy a competitive edge in the international market.

The second apprehension expressed by the detractors of GATT is that India has to import at least 3 per cent of the domestic demand for agricultural products. This will have to be examined in broader context than merely as an imposition to force that country to spend its precious little foreign exchange on food imports which might otherwise have not been required. Considering the vagaries of weather and other crop uncertainties, it might prove to be a desirable thing to import a small amount of agricultural products from abroad. The country could as well escape the operation of this clause under the pretext that it is a developing country reeling under weak and unfavourable balance of payments position.

The third area of concern is that India has to abandon its policies on public procurement operations and public distribution system (PDS). The question of undue influence on the Public Distribution System does not arise because the PDS is meant to help the poor consumer and not the farmer. Logically, the dominant agricultural exporting countries would like to see India distribute its food under PDS. By so doing, the demand for food will increase and India has to import food from outside.

Serious apprehensions are also expressed about the effects of patenting seeds and genetic materials. It is admitted that as a result of patenting seeds, the food price will surely rise, but the farmers will not be affected as much as they proclaim it to be. The real fact regarding the patenting of seeds is that our farmers have the choice to buy high productive seeds if they want an increase in their agricultural production.

Thus, if they want a higher yield, they can go for a good variety of seeds, and additional cost will be compensated as a result of increase in the production. Farmers will have the privilege to use seeds saved during particular year for the subsequent years as long as they remain grain producers. Similarly, researchers are free to develop a new variety of seed by using one protected variety. This might also induce more original research in the development of new crop gene/seeds suiting not only the Indian agroclimatic conditions but that of other developing countries as well and thus offer an opportunity to the country to export these and earn royalties.

The above analysis shows that India will stand to gain in some fields while simultaneously it will have to exercise caution so as to capitalise more on the advantages and at the same time evolve some new ways to combat effectively the areas which are detrimental to India's interests.

Agriculture obviously is an important sector of India's economy and any distortions in its development are likely to produce far reaching implications for the people, in general, and farmers, in particular. At no stage can the country afford to ignore the economic and social effects of agriculture. But in the changed world economic scenario, it may not be possible or even desirable to keep agriculture as a highly protected sector of the economy. There is need to accept the liberalised agricultural trade and orient the country's policy towards that end.

This will call for all round improvement in agricultural efficiency to bring it closer to the international norms. While as in absolute terms, the country might be one of the largest producing countries of agricultural products, in relative terms, the productivity levels are still very low and need to be enhanced substantially. A few hard facts may make it more clear. In rice, India's rank in the world production is 2 but its rank in yield per hectare is 28, in wheat, the relevant figures are 3 and 31, in milk, 1 and 34, in groundnuts, 1 and 34 and in Jute and fibres, 1 and 11.

Agriculture related provisions in GATT also offer ample opportunity to India to expand its agriculture export market. However, it cannot do so until it is able to produce more than its" domestic consumption. This can be achieved only by quick increase in the agricultural production. Thus, we should address more to modernising the agricultural sector and providing farmers the latest developed agricultural equipments. At the same time we should concentrate more on research and development in this sector, so as to develop more productive hybrid seeds, so that our farmers may not fall prey to the multinational corporations.

Conclusion

An interesting international debate is going on over the GATT recommendations. In some sectors, it is held that these recommendations are a consolidations of the economic dominance of the developed countries like USA, EC, Japan, etc. and are anti-third world; while others are hopeful that it would cause higher employment generation, more investment and more trade and would ultimately bring discipline in trade practices.

The Scenario that would unfold can not be predicted now as there is a time span of 10 years on an average for the implementation of GATT recommendations. The developing countries could use the phase-out time to readjust their development priorities and evolve strategies to farmer sufficient benefits from the liberalised world regime. The global equations and balances of power are hinging on pendulum and third world countries will be required to work hard to swing it in their direction.

11

Crops for Commercial Purposes

Tobacco

FCV tobacco NLS 4 with an yield potential of 2,500 kg/ha was found for light soils of Andhra Pradesh and Karnataka and KST 19 Thrupthi (1,500/kg/ha), resistant to black shank, root-knot nematode and tolerant to drought stress, for light sorls of Karnataka are suitable for cultivation.

Nine direct interspecific hybrids 4 bridge-cross hybrids have been produced, while incorporating aphid resistance in FCV cultivars. Among these, hybrids ((Nicotiana numbratica-mesophial), Nicotiana tabacum), {(Nicotiana benthamiana-repanda), Nicotiana tabacum), (Nicotiana tabacum, (Nicotiana gossei-excelsior) and (Nicotiana tabacum, (Nicotiana gossei-gulutinosa)} were produced for the first time. All F interspecific hybrids were screened for aphid resistance and 6 showed resistance. Among 4 bridge-cross hybrids, 2 with Nicotiana gossei, as a donor parent, have been found resistant to aphids.

Among various crops tested for alternate crops to tobacco in black cotton soils, growing of hybrid maize during kharif, followed by chickpea in rabi, showed a benefit : cost ratio of I : 4.40, followed by tobacco monocropping with a benefit : cost ratio of I : 1.83.

Sunhemp green manuring (kharif) + bidi tobacco (early rabi) followed by maize during summer gave the highest gross income.

Maximum net returns of Rs. 64,933/ha in chewing tobacco were recorded in Bihar with a combination of 25 per cent nitrogen as mustard-cake + 25 per cent nitrogen as castor-cake + 50 per cent nitrogen as urea.

Crop Protection. Sequential sprays of Bt kurstaki 1.0 kg/ha and SINPV 250 larval equivalent (LE) / ha effectively controlled Spodoptera litura in tobacco nursery and gave a benefit : cost ratio of 1 : 1.35 in the chemical control practice; adopted by nursery-growers.

Score 23 per cent EC (Difenoconzole), a systematic triazole fungicide, is identified as a better chemical than recommended Bavistin 50 per cent wp for controlling anthracnose in FCV tobacco nurseries.

Brown-spot damage in the field crop can be effectively checked by 3 or 4 foliar sprays of systematic fungicides Tilt and Score at 0.2 per cent or non-systematic fungicides Indofil M-45 and Foltaf at 0.5 per cent in light soils.

BIMP practices for Helicoverpa armigera which included HaNPV at 450 LE/ha and management practices to improve natural enemy activity, through erection of bird perches and raising 2 rows of Tagetes around FCV tobacco fields, resulted in only 1.30 per cent damage by Helicoverpa armigera as compared to 20-30 per cent in farmers' method.

Seed Production

Breeder Seed Production

A total of 2,643.6 tonnes of breeder seed has been produced during the indent of 2,501.6 tonnes. Major quantities belonged to cereals, oilseeds and pulses.

Isolation distance for castor may be increased upto 300 m for certified seed and the present isolation distance of

300 m for foundation seed should be increased to 600 in for maintaining proper genetic purity in castor by modified method, which throws only 2-3 per cent monoecious plants.

Distinguishing characteristics of 68 rice varieties pearl millet hybrids GHPB 263, 316, 183 and 229) and their parental lines cotton varieties and 2 new sunflower hybrids (PSFH 67, PKV SH 27) and their parental lines have been compiled.

Laboratory techniques have been standardised for identification and characterisation of varieties, hybrids and their parental lines using electrophoresis profiles polyacrylamide gel electrophoresis (PAGE) patterns of soluble proteins of cotton, rice, sorghum and soyabean.

Other distinguishing varieties were a major problem in seed production of soyabean. The identified other distinguishing varieties are true off-types and not a weathering effect. Off-type plants were characterised by foliage colour, growth habit, pigmentation, flower colour, hairiness, etc.

In rice recommended dose of NPK + 25 kg $ZnSO_4$/ha as basal along with 2 foliar sprays of $ZnSO_4$ (0.5%) at boot leaf and at panicle initiation proved better for enhancing pollen production and pollen viability in restorer parent for good seed-setting and seed yield.

In wheat, use of 20 per cent less seed than recommended (80 kg/ha) and normal dose of fertilizer (120 kg N+60 kg P_20,s/ha) gave good seed yield and graded seed quantity as well as quality. For seed production in mungbean and urdbean, normal sowing in July with normal spacing at 30 cm has been found better.

Polythene-lined cloth-bag or gunny bag in general was superior to ordinary cloth or gunny bag, in maintaining higher germination and vigour of seed during ambient storage. In soyabean more than 70 per cent of germination could be maintained for up to 25 months in the polythene-lined cloth-bag, as compared to 14 months in the ordinary cloth-bag.

Good quality sorghum (cv. M 35-1), sunflower (cv Morden) and soyabean seeds with high initial germination and vigour could be retained at a desired level of germination up to 9 in ambient conditions having moderate RH and temperature.

Hydration of seed with GA3 and simple hydration-dehydration with or without 'Thiram' dry-dressing, increased the speed and final field emergence in maize, sunflower, mustard and onion. These treatments also increased seed yield than recommended Bavistin 50 per cent wp for controlling anthracnose in FCV tobacco nurseries.

Rice-bunt incidence in traditional rice-growing areas of Karnataka, Tamil Nadu, Andhra Pradesh and Orissa, either absent or were very low compared to very high bunt incidence in non-traditional rice-growing states Haryana, Punjab, Himachal Pradesh and Madhya Pradesh.

The seed infection of bacterial leaf-blight in the harvest of all plots of rice raised from streptocycline' seed treatment was only 5 per cent compared to 85 per cent in untreated-plot seeds at Bhubaneshwar. This indicates that seed-borne Xanthomonas oryzae pv. oryzae can be taken care through 'streptocycline' seed treatment, Seed-borne nature of the urdbean leaf crinkle virus was no there is correlation between seed-borne percentage and secondary spread in fields; as this is influenced by crop type, varieties and time of sowing. In cauliflower, Alternaria blight disease reduction and seed yield increase were recorded in plots sprayed with Contaf (0.05%), followed by Rovral (0.2%) and Dithane M-45 (0.2%)

Seed recovery percentage decrease and seed rejection increased in general with increase in sieve aperture size. Hence aperture sizes recommended are as follows : mungbean - 2.8 mm round; urdbean 3.2 mm oblong; soyabean - 3.6 mm round; sunflower - 2.3 mm oblong; safflower - 2.4 mm oblong; sunflower parental lines CMS 2A an CMS 2B - 1.6 mm oblong; AKR I - 1.4 mm oblong and sunflower hybrid PSFH 67 - 1.4 mm oblong (at 0.6 kg/min, feed rate and 800 strokes/min). In soyabean, mechanical Thrashing at 720 rpm showed lowest germination (65.8%) associated with highest mechanical Thrashing at 500 rpm k7.9 rpm tip speed) compared well in germination (81.8% and 80.6%). Toothless concave resulted in acceptable seed quality at the cost of excessive seed loss in chaff.

In safflower, harvesting losses and mechanical damage were observed at 4-5 per cent and 5-9 per cent. Thrashing and cleaning efficiencies in mungbean favoured cylinder speed of 6.41 m/sec concave clearance of 10 mm (front) and 7 mm (rear). Storage indicated that effect of drying methods manifest in different performances in storage. Sun-drier proved the worst storer while forced air drier was the best storer.

Trichoderma harzianum isolate PDBCTH 2 an Gliocladium virens have proved promising in inhibiting mycelial growth of root-rot fungus Sclerotium rolfsii. Gliocladium virens + carboxy methyl cellulose formulation has found most effective against chickpea wilt and root-rot completes. It significantly improved seedling emergence, plant stand and grain yield of chickpea when used alone or with Vitavax. Trichoderma harzianum (PDBCTH 2, Trichoderma koningii, Gliocladium virens and Gliocladium deliquesens are found effective against Meloidogyne incognital maximum mortality of 97 per cent was by Trichoderma koningii.

Entomophilic nematodes Steinernema isolates from Devanahalli (Karnataka) killed Plutella xylostell and Opisina arenosella larvae within a day of inoculation and Helicoverpa armigera, Spodoptera litura and Corcyra cephalonica within 2 days.

The IPM module for cotton crop developed by the NCIPM, which performed well on farmers' field, was validated on 200 involving 124 cotton-growers. It comprised planting a row of maize with cowpea as interplants all around the field to enhance population of friendly insects (predators and parasites), planting of Setaria in between 9 and 10 cotton rows to attract insect predatory birds, seed treatment with Imidacloprid at 7 g a.i./kg of cotton-seed, 2 releases of parasitoids Trichogramma chilonis at 150, 000 at egg-laying of Helicoverpa, and application of HaNPV at 250 LE/ha and 2-3 of home made NSKE (neem seed-keranel extract) in the basis of ETL.

On the basis of the 15 years data on adult-moth catches from, adjoining districts of Hyderabad and also rainfall-distribution pattern over these years, a preliminary model has been formulated to predict Helicoverpa armigera outbreaks. Deficit monsoon (June to September) and surplus rainfall in November can lead to an outbreak of the pest.

Apiculture

Mass queen rearing studies in Apis mellifera showed significantly higher larval acceptance with a queen cell-cup of 9-mm diameter (77-5%). Acceptance was high (9017() when cell-cups were primed with royal jelly. Young larvae of less than 24-hr or up to 36-hr were found better for grafting. These gave better emergence and quality of queens as judged by weight of queens which ranged 165-184 mg in 8-9 mm diaqueen-cups, 138-144 mg in cells prime with royal jelly and 125-138 mg when 20 larval grafts were implanted in 10-frame queenless colony highest brood development (increase 450 q 30 cm). Hence increasing entrance breadth up to 50 per cent over control is effective in reducing temperature and relative humidity in summer.

Bee-Ecotypes

Honeybee populations (Apis cerana indica) of high-ranges, highlands, and mid-lands of Kerala were studied with reference to morphometric characters. The bees from high-range were distinct with longer probosics, antennae, wings and legs. Multivariate analysis of the morphometric data revealed presence of 4 clusters/ecotypes in populations; all showing susceptibility to TSBV (Thai Sacbrood Virus).

Bee Pollination

Pattern movement of Apis mellifera foragers between male-fertile (male) MS-234 B and male strile MS 234-A (female) lines of sunflower sown in the ratio of I : 30. It was noticed that 28.32 to 34.56 per cent (mean 31.59%) movement of foragers was from male to male flowers and 30.47 to 37.79 per cent (mean 33.51%) from female to female flowers, and 16.83 to 18.63 per cent (mean 17.55%) bee visits were from male to female flowers and 18.14 to 20.03 per cent mean from 19.00%) from female to male flowers. The mean intersexual flower visiting movements varied from 17.55 to 19 per cent which is considered useful for hybrid seed production.

Apis mellifera was the dominant forager in Shahi variety of litchi, followed by Apis dorsata, Apis cerana indica and Apis florea. Apis mellifera have been visiting litchi flowers ranged from 38.60 to 54.89 per cent (mean 46.%), followed by Apis dorsata indica from 15.01 to 29.16 per cent (mean 22.56%) and Apis florea from 9.68 to 27.11 per cent (mean 15.84%). Fruit set by 78.66 and 55.12 per cent with open to all pollinators and caged without bees. In sapota, fruit set was 24.85 per cent and 28.15 per cent at 20-m and 40-m distances from bee-colonies. Bees activity and fruit setting were lower at 70-m and 100-m distances. Two-bee colonies/ha were required for optimum pollination of Cricket Ball cultivar of sapota.

Honey Storage

Effect of storage temperatures of 3°-4°C room temperature, and 40°C was studied on unifoloral melliferous honeys of Brassica campestris var. toria and Brassica campestris var. sarson stored in containers of glass transparent plastic, opaque plastic, tin, stainless steel and hindolium. After 6 months, moisture content from original 21.6 per cent ranged between 21.7 and 21.8 per cent in toria honey. A decrease in sucrose content was observed from 0.84 per cent to 0.73 - 0.75 per cent at 4°C.

Fiche's test was positive at 40⁰C only in transparent plastic, tin and steel containers. Acidity (at 6 months) also increased from original 0.131 - 0.136 to 0.143 per cent at 40°C in all containers. Colour, aroma and flavour (Organoleptic evaluation) were several affected at 40°C in tin container after 3 months.

They were severely affected after 6 months in tin containers at all temperatures and in transparent-plastic bottles and steel containers only at 40°C. After 3 months, a uniform hard granulation with fine crystals was observed; this was 100 per cent at refrigerator temperature and 10-20 per cent at room temperature in hindolium containers. Darkening of honey was observed in tin containers both at room temperature and 40°C and resulted indeterioration of flavour at 40°C after 3 months. Honey granulation in refrigeration was negligible in hindolium and opaque plastic jars but in others, it varied from 20 to 75 per cent.

Bee Diseases

As yet bacterial diseases have not been reported from Apis mellifera in India. Incidentally, a bacterial disease identical to European foulbrood was detected in Sirmor (Himachal Pradesh), which caused 30-40 per cent mortality of Apis mellifera colonies. The bacterium is highly sensitive to oxytertracyclin.

Bee Predators

For controlling wasps predating on honeybees, fish-bait technique has been found very effective attracting as many as 23.75 predatory wasps/hr. Meat bait proved second in effectiveness and attracted 16.25 wasps/ha.

Rodents

Lesser banicoot rat Bandicota bengalensis continues to dominate pest complex in irrigated cropping systems of Punjab, Madhya Pradesh, Gujarat, Andhra Pradesh, Karnataka,

Himachal Pradesh and northeastern hills region, besides Milardia meltada, Tatera indica and Mus sp. Common house rat Rattus rattus was reported from chickpea, wheat and soyabean in Madhya Pradesh. Dryland crops pearl millet, sorghum, maize and pulses were found predominantly infested by Milardia meltada and Tatera indica in Andhra Pradesh, Karnataka and Rajasthan, Short-tailed mole rat, Nesokia indica, was found in canal irrigated cotton fields in Bikaner (Rajasthan). This species is a new introduction on arid regions. Earlier, N. indica was trapped from arid afforestation plantations in Nagaur district.

Larger bandicoot Bandicota indica nemorivega was reported from rice fields of Umiam areas in Meghalaya. In Punjab, sugarcane and rice-wheat rotations predominantly harboured Bandicota bengalensis, and groundnut, pulses, millet and cotton recorded predominance of Tatera indica in Punjab. The house mouse Mus musculus has successfully carved its niche along with Bandicota bengalensis in orchards, nurseries, vegetable crop fields and tea plantations in Himachal Pradesh.

In western Rajasthan, gerbils (Meriones hurrianae and Tatera indica) inflicted more than 50 per cent damage to sole crops of mungbean and mothbean at the vegetative stage. Infestation was reduced (38.3%) when they were intercropped with pearl millet. During rabi, wheat, chickpea and vegetables suffered 5.0 to 17.5 per cent damage due to rodents. In Churu and Bikaner (Rajasthan), groundnut suffered 30-65 per cent yield losses to merion gerbils. Damage to wheat was 2.3-8.7 per cent in Madhya Pradesh; 4.0-6.0 per cent in Punjab; 4.4-14.8 per cent in Gujarat. Similarly rice suffered estimated yield losses of 0.7-12.9 per cent in Madhya Pradesh; 4.0-6.0 per cent in Punjab and 4.9-6.5 per cent in Madhya Pradesh and extent of damage to peas was 5.3-7.8 per cent in Himachal Pradesh. In pearl millet, rodents caused 3.5-4.2 per cent earhead damage, and sugarcane suffered 4.0- 10 per cent losses in Gujarat. Tomato, cauliflower and cucurbits recorded 5.2-9.7 per cent, 2.7-7.5 per cent and 5.2-18.4 per cent damages.

Burrowing pattern of Bandicota bengalensis in Meghalaya and Nesokia indica in Rajasthan showed an average of 4-15 openings with 1-6 brood/nesting chambers/bur row. Latter size of Bandicota bengalensis ranged from 3 to 9 in Meghalaya and was 6 for N. indica in Rajasthan. Mole-rats in Meghalaya, hoarded 2.6 kg of foodgrains/ha and it was 1.5-1.7 kg/burrow in Himachal Pradesh. In Rajasthan hoarding as high as 2.5 kg of groundnut-kernels was found for the first time in each burrow of Meriones hurrianae. A germ-cell mutagen, ethyl methane sulphonate, at 100-150 mg/kg b.w. and 500 mg/kg can cause dose dependent sperm abnormalities in Bandicota bengalensis and Rattus rattus.

Difethialone, a third generation anticoagulant rodenticide, caused cent per cent, mortality of test species at 25 ppm in locally available baits. In field mortality was observed between 3 and 10 days of its consumption and efficacy was at par with Bromadiolone (0.005%) in loose/waxblock baits. Cholecalciferol, a vitamin 133-based rodenticide, has also shown its efficacy against pest rodents at 0.075 per cent concentration in freshly prepared baits. Zinc phosphide (2%) bait, followed by Bromadiolone (0.005%) baiting have been recommended for rodent management in arable crops. In Gujarat, two applications of Bromadiolone (0.005%) at 10 g per burrow, first at sowing and second at pod formation stage were recommended for groundnut.

Social engineering activity on rodent control was undertaken by all cooperating centres for technology assessment and refinement at farmers, fields. Adapted villages had a success of over 75 per cent.

Wheat Sonalika and WJ 64 found fairly tolerant to brown mite Petrobia latens and C 306 and WJ 141 found susceptible. Amblyseius alstoniae was identified as a potential predator of wheat-mite. Ethion at 0.05 per cent minimised mite damage, followed by Phosphamidon (0.03%) and Monocrotophos (0.03%).

Pigeonpea was damaged by sterility mosaic to the extent of 80 per cent in Valsad, 30-84 per cent in Surat and 10-20 per cent in Bharuch in Gujarat. Mite vector Aceria cajani population was highest (6-8/leaf) on 180-200 days old crop at the reproductive phase. Disease incidence was maximum on July-sown crop and mite vector was abundant in December-sown crop. At Bangalore, Maruti recorded maximum sterility (29%) incidence, followed by GPS 7 (22%) and GPS 26-C (16%).

Maruti also harboured a maximum of 24 mites/leaf. Sujatha 1-2, S-1-3, GC 11-39, GS 1, GPS 42, BDN 2, GS 1, Maruti though recorded disease incidences of 2 per cent, did not favour vector population build-up. At Bharuch (Gujarat), BP 97-2, and BT 20 were free from sterility disease as against 30-39 per cent disease recorded in susceptibles BDN 2, BP 97-4 and G 7. At Bangalore, survival (4-5/leaf) of Aceria cajani during off-season (December to April) on weed Atylosia scarbaeoides was important source for carry-over of mite to succeeding cropping on stubbles/ratoons/voluntary/perennial pigeonpea in May-June.

In Gujarat, Sorghum mite Oligonychus indicus which preferred lower leaves reached peak in October. This was associated with the peak of Stehorus pauperculus. Lines SR 389, CSH 5, IS 2312, GSH I showed lower mite infestation compared to 60 per cent leaf infestation observed in CSH I and GJ 36.

Tetranychus cinnabarinus infested brinjal almost throughout the year at Varanasi with a peak of 116 mites per 2.5-cm 2 leaf area during second fortnight of March when temperature and relative humidity were 22°C and 38 per cent. Predatory phytoseed mites were abundant in November (35/leaf) at 23°C and RH of 52 per cent. Okra mite attained peak in second for night of August on Pusa Sawani at Ludhiana. Peak mite incidence at Varanasi was associated with 35°C and

39 per cent RH in the first fortnight of May, and predators Amblyscius longispinosus and Amblyseius multidentatus were also maximum during this.

At Pusa, a maximum of 51 mites/2.5-cm leaf area was observed during second week of June when maximum and minimum temperatures were 41°C and 27°C, and all weather factors together accounted for 84 per cent variation in mite population.

Chilli mite Polyphagotarsonemus latus reduced fruit yield in G 4 chilli by 40 per cent, inflicting a monetary loss Rs 4,950/ha in Gujarat, and in West Bengal 35 per cent loss was recorded on Suryamukhi. In West Bengal, this mite also infested jute, cowpea, balsam, tea, marigold, datura, mungbean and many weed plants. On chillies, mite recorded peak a, 29°-33°C and RH of 65-93 per cent. PS 8 was highly susceptible and Sundarai was fairly tolerant to mite damage.

Large-scale damage to coconut-nuts is by an eriophyid mite Aceria guerrerronis; causing premature dropping of young nuts of buttons, and imparting discoloration or scars in the surface of older nuts. The damage was more in Bangalore and adjoining areas bordering Tamil Nadu. The mites congregate beneath tepal, damage soft tissues, that later turn to necrotic lesions and longitudinal streaks, resulting in splitting and malformation as nuts grow.

At Varanasi, Ethion (0.05%) controlled brinjal mite, and at Pusa use of wettable sulphur and Dicofol realised benefit : cost ratio of I : 5.8 to I : 6.1 in okra. Repeated applications of synthetic pyrethroids against okra fruit-borer caused resurgence of mite in Punjab and Ludhiana by eliminating phytoseid mite predators and these insecticides were also observed to induce egg-laying in Pusa, at Bangalore, rose-mite was effectively managed by spraying. Avermectin, Profenophos, and a combination of Monocrotophos or Dicofol with pongamia and mahua oils.

Different fractions (powder, oil and oleoresin) of turmeric and garlic showed effectiveness against Tyrophagous putrescentiae and Suidasia, nesbitti infesting stored wheat in Haryana. Aamblyseius longispinosus (a phytoseiid predator) released at a predator: prey ratio of 1:300 against Tetranychus urticae infesting rose in polyhouses at Bangalore could bring down prey population within 15 days compared to release ratios of 1 : 450 and 1 : 900.

Pesticide Residues

In multilocation trials on rice, basal application of granules of Phorate (1.2-2.4 kg a.i./ha) and Carboufuran (I kg a.i./ha) resulted in either no residues or residues below maximum residue limits in grain and straw at harvest. However, a waiting period of 60 days is suggested for Lindane on rice to ensure safety to consumers. On wheat, application of chloropyriphos as seed treatment (0.9-g a.i./kg seed) and as foliar spray (400 g ai/ha) at earhead formation stage showed no residues in grain and straw at harvest, though seed treatment resulted in low levels of residues in soil.

In sugarcane, HCH application at 2.4 kg ai/ha as soil drenching resulted in very little HCH isomers in cane and juice. Application of Mancozeb (1.2-2.4 kg as foliar spray), Phorate (2.5-5.0 kg/ha as soil application), Quinalphos and Chlorpyriphos (25, 50 ml each/kg seed) on groundnut resulted in no residues in kernel, shell and soil at harvest. Application of Iprodione on mustard at 500 g a.i./ha at 50 per cent pod formation stage showed no residues in grains at harvest.

On soyabean, a waiting period of 2-3 days was recommended for Lindane (100, 500 g a.i./ha) and 4-5 days for Phorate (2,4 kg a.i./ha), following application at flowering. On okra, safe-waiting period of 4-6 days was recommended for Lindane (at 0.5, 1 kg a.i./ha), 3-4 days for Chloropyriphos (250, 500 g a.i./ha) and 1-2 days for Fluvalinate (50,100-g a.i./ha), applied at fruiting, indues below maximum residues limit were found in cabbage on application of Quinalphos (625 g/ha), and cypermethrin (125 g\ha) at head formation stage even on zero day.

Mancozeb at 1-2 kg a.i./ha and Carbendazim at 0.5-1.0 kg a.i./ha on chillies required safe-waiting period of 3 days was recommended for Antrakcol (2-4 kg a.i./ha) and Dichlorvos (0.5-1.0 kg a.i./ha) at fruit-development stage. However, Mancozeb (0.75, 1.5 kg a.i./ha) and Propineb (0.875, 1.75 kg a.i./ha) on tomato at fruiting stage resulted in low residues.

On appale, Dichlorvos (0.5 and 1.0 kg a.i./ha) applied at fruit-development stage needed a waiting period of 8 days, and Antracol (2.5, 5.0 kg a.i./ha) and Lindane (0.5, 1.0 kg/ha) showed no residues or below maximum residues limit values.

Cumulative effect of wrapping + ribbon in farmers' fields in on maize showed higher yield (2,065 kg/ha), compared to control (764 kg/ha) Tobacco-leaf decoction (10%) spray on sorghum at milky stage reduced bird damage at Anand (Gujarat). In Punjab, sorghum as screen around maize yields reduced *parkeet* damage to ripening maize.

The thorn variety passifera of oilpalm found bird resistant. The farmers were advised to protect crop by traditional methods like covering with gunny bags or partly with leaf baskets and dried male inflorescence of oilpalm.

In Punjab bird damage caused to sunflower was assessed by selecting ten fields at each location. Extent of grain damage ranged from 0 to 32 per cent of total surface area with an average dama- of 5.84 q 12.20 per cent. In safflower, peripheral plants were more prone to bird damage than the inner plants.

Among 14, the most beneficial birds are Cattle Egrets, Rosy Pastors, Indian Myna, Bank Myna and Black Drango which controlled Helicoverpa armigera on chickpea by feeding on its larvae. Two sprays of Endosulfan (0.07%) at pod-formation stage coupled with bird activity checked pest population.

The prey-searching efficiency of insectivorous birds was better in chickpea (var. ICCC 4) grown at 60-cm inter-row distance compared to 45-cm inter-row distance, in lucerne, Cattle Egret was most abundant, Feeding on insect-pests, particularly Helicoverpa armigera (49.4%) followed by Bank (22.5%), and Indian Myna (8%).

Cutting practices of lucerene removed about 50 per cent of Helicoverpa larvae and out of remaining, nearly 91.77 per cent were reduced by predatory birds, following labourers engaged in cutting, resulting in a total check on pest.

In Rajasthan Cattle Egrets regulated larval population of saw-fly Athalia proxima on mustard to an extent of 60 per cent after irrigation, and brought down population much below economic threshold level. The birds picked profusely sawfly maggots up to 8-10 has in a day, with a picking success from 48.45 to 65 per cent.

Green bee-eater Merops orientalis gut analysis at Ludhiana revealed that Hymenopteran insects (almost exclusively honey bees) constituted 81 per cent of total food content. Other insects were Odonata (11.62 %), Diptera (2.4 %) , Lepidoptera (1.8 %), Orthoptera (1.27%) , Coleoptera 1 per cent and others (0.93%) throughout the year, exception October (31.5 %). These studies conclude that bee-cater is a pest on honeybees and warrants suitable control measures.

Whitegrub Management

Quinalphos, Chlorpyriphos, Diazinon at 5 g a.i./kg seed and Imidacloprid at 2.8 g a.i./kg seed were effective as seed dresser in groundnut, giving 73 to 93 per cent mortality of first instar grubs exposed to soil surrounding treated seeds, 10 days after sowing. Quinalphos and Chlorpyriphos persisted in soil for a longer period giving 73 and 67 per cent mortality of first-instar grubs 15 days after sowing. Among granular insecticides, seed furrow application of Isozophos, followed by Fipronil were effective. Presently recommended Phorate 10G at 2.5 kg a.i./ha yielded 1.913 tonnes/ha.

Fusarimum moniliforme, F. chlamydospores, F. oxysporum, Aspergillus niger, A.chevalieri Trichoderma viride, Penicillin sp. and Rhizoctonia bataticola were pathogenic to whitegrub, Holotrichia consangiomea. Metarhizium anisopliae, isolate Ma-4 from Durgapura proved highly virulent to Holotrichia consanguinea causing 100 per cent mortality within 6 days of incubation at 27°C.

Nematode Management

Root-knot nematode Meloidogyne incongnita and cyst nematode Heterodera cajani have recorded as major nematode pests causing appreciable yield losses to mungbean, pigeonpea, chickpea and cowpea in Bihar, Uttar Pradesh, Maharashtra, Tamil Nadu and Haryana. Wilt disease incidence in chickpea was associated with combined infestations of Fusarium oxysporum F. sp, ciceri and root-knot nematode in Uttar Pradesh and Madhya Pradesh.

Rice-stem nematode Ditylenchus angustus caused "Ufra" disease in most deep-water rice-growing areas of Assam. This nematode was also associated with sheath rot, Sarocladium oryzae, leaf blast and neck blast Pyriulia gricea.

ML 3, ML 62, ML 10, and K 851 mungbean and GAU 1 and 82 lb cowpea were found resistant to Meloidogyne incognita.

Infestation of root-knot nematode and cyst nematode on mungbean, pigeonpea, chickpea and cowpea could be successfully minimised by seed treatment with Carbosulforn (25 ST) at 3 per cent w/w or with Monocrotophos at 0.1 per cent.

In wilt of chickpea, the treatment of Carbosulfone (25 ST) at 3 per cent w/w + Bavistin at 0.2 per cent was found effective.

Ufra disease of rice can be managed with resistant 'Rayada'-116-06 treated with Carbosulfone (25 ST) at 3 per cent w/w, followed by foliar spray of Carbosulfone at 0.2 per cent at 40 and 120 days after sowing. Combination of seed treatment and foliar application was effective and economical against rice-stem nematode.

San Jose scale Quadraspidiotus perniociosus at Solan, Shimla and Kullu was parasitised by Aphytis sp. (proclia group) and Encarsia perniciosi and predated by Chilocorus bijugus grubs. Aphelinus mail was active during May-December and parasitised apple woolly aphid Eriosoma lanigerum, in spite of its low population this year.

Trichogramma chilonis parasitised 29 per cent of Helicoverpa armigera eggs on tomato, and larval parasitisation by Campoletis chlorideae was up to 12.5 per cent at Solan. Peal leafminer Chromatomyia horticola ,was parasitised by Diglyphus sp. (Eulophidae) on pea (21.1-58%), broccoli (0-15.4%) and sweet-pea (0-100%) at Solan. Trials with Trichogramma pretiosum and HaNPV on tomato fruit-borer resulted in 50'(egg parasitism by Trichogramma pretiosum and 30-50 per cent infection by HaNPV at Bangalore.

Apanteles sp. and Bracon sp. have been found parasitising foliage-feeding Phthorimaea operculella (PTM) larvae. Two Hymenopteran parasitoids and a coccinellid predater were recorded on PTM larvae in stored potatoes. Miniature arnies (20-kg capacity) were constructed to simulate village sorage for potato and testing bilogical control agents. Bt at potato and for testing bilogical control agents. Bt at 1.0 g/kg tubers was most effective in reducing tuber Chelnus blackburni at 50,000 adults/ha and Chelnus blackburni at 50,000 adults/ha were equally effective in reducing leaf mines by PTM, SINPV at 750 LE/ha (=4.5, 1012 polyhedral bodies/ha) was effective, causing 87.50 per cent larval mortality of Spodoptera litura and gave 18.45 tonnes/ha tuber yield at Pune.

Two exotic weevils Neochetina eichorniac and Necohetina bruchi controlled water hyacinth in Disangmukh area of Sibsagar (Assam). Floering was also has in the remaining water-hyacinth areas of Sibsagar.

The pyrailid Bactra venosana damaged Cyperus rotundus severely in pot trials conducted at Bangalore. Larval feeding caused reduction in plant growth, tuber weight and regeneration capacity of tubers.

In Himachal Pradesh, Zygogramma biclorata completes 5 generations in a year. In field, maximum adult population (1.3 beetles/plant) was observed during June end and first week of September and middle of October. In Punjab, this beetle is widespread on parthenium in Ripar, Nawanshahr and Jalandhar.

Cotton

Cotton CNH 154, CNH 108. CNH 1012, CNH 123. RS 213. RS 2106, HD 328, and HD 368 are identified as resistant to cotton leaf curl virus. Intra-hirsutum hybrid CINHH 109 is performing well in Gujarat, Maharashtra and Madhya Pradesh and CSHH 88 in Punjab, Haryana and Rajasthan. Advance cultures,viz. ES 12, ES 12 and HS 6 with increased oil content (22-25%) and tolerance to jassids have showed consistently good performance Gujarat, Maharashtra and Chhattisgarh and Madhya Pradesh.

Paired-row planting and liquid-fertilizer through drip irrigation resulted in higher seed-cotton yield and higher water-use efficiency; saved 50 per cent of irrigation water. Drought-tolerant genotypes NA 1589, CPA 423, SUPRA 2, AH 140, TSH 188, G. Cot by 8, G. Cot DH 9 and G. Cot 17 have been identified on the basis of the physiological and biochemical attributes. At Coimbatore, photosynthetic rate, nitrate reductase activity and leaf chlorophyll content of cotton grown in elevated CO_2 atmosphere were higher than control. At Nagpur, preparation of ridge furrow at the first interculture increased soil moisture at 0-30 cm and 30-60-cm depth at peak boll development stage from 14.91 and 21.73 per cent in control to 16.09 and 22.51 per cent. This recorded significantly higher cotton yield.

At Khandwa (Madhya Pradesh), Cotton + groundnut provided maximum seed-cotton yield, and cotton + greengram (mungbean) and cotton + blackgram (urdbean) were found profitable in Vidarbha. In Andhra Pradesh, cotton-chillies-cotton was found better with handsome return of amount.

Plant Protection. Rapid and inexpensive pyrethroid resistance detection unit has been developed at Nagpur for detection and determination of resistance frequency in field strains of Helicoverpa armigera.

Seed treatment with Imidacloprid (70 ws) at 5 g/kg of seed was effective in controlling sucking pests up to 70 days after sowing. The IPM module comprising tolerant variety, seed treatment, release of parasites, balanced fertilizers, using soft-and-selective insecticides, and hand-picking of eggs, larvae and damaged fruiting parts proved superior to traditional spray schedule. Three releases of Chrysopa cornea (10,000/ha/week) and 5 releases of Trichogramma chilonis (250,000/ha) between 45 and 110 days were found effective in controlling bollworm. Sirene, a new formulation of pheromone, effectively checked the pink bollworm infestation. Cowpea intercropped with cotton resulted in lesser aphid and leafhopper population and lesser bollworm incidence than cotton alone.

Maize, Cassia occidentalis, castor, sunhemp and marigold, when intercropped with cotton or found growing in the fields, harboured parasitoids/predators of cotton-pests at Anand, Gujarat.

Biocontrol-based IPM damage. Parasitisation by Trichogramma chilonis on Helicoverpa armigera eggs was highest in Punjab. BIPM practices adopted at Hyderabad which included intercropping with groundnut, excelled included intercropping with of cotton pests. Incremental cost : benefit in BIPM was high (10.07) as compared to farmers practice (1.55) and judicious usage of insecticides (1.59).

Sugarcane

Test entries Co 91011 and Co 91005 gave best performance at Anakapalle for cane yield (118.7 tonnes/ha and

117.8 tonnes/ha) when compared with Tamil Nadu, Andhra Pradesh and Orissa. In western Uttar Pradesh, Rajasthan, Punjab and Haryana, CoS 91269 was best at a number of locations, Co 91010 gave the highest cane yield of plant crop (1·19.5 tonnes/ha) and ratoon-crop (98.06 tonnes/ha) at Coimbatore.

Ninety intraspecific hybrids of Saccharum robustum showed a mean sucrose content of 10.9 per cent, which is substantially higher than sucrose content of the natural population of Saccharum robustum. Maximum sucrose content was 14.07 per cent. Thirteen clones, viz. Co 98001 to Co 98013 have been identified as promising on the basis of yield and quality. The cane yield and sucrose of these ranged from 95.69 per cent tonnes/ha to 141.75 tonnes/ha and 18.68 per cent to 21.37 per cent. Co 98004 combined high cane yield of 125.27 tonnes/ha with sucrose percentage of 20.24. New clones of Saccharum spontaneum and Erianthus from northeastern region in crosses with hybrid cane varieties gave hybrids with better vigour and tillering capacity.

The application of 20 tonnes of farmyard manure (FYM) or cane trash + sulphitation press-mud each at 10 tonnes/ha, along with recommended dose of NPK sustained productivity of sugarcane at the higher level than NPK alone at the recommended dose. The application of 100 and 125 per cent of recommended dose of N (150 kg/ha) in 2-3 equal splits increased yield of seed-cane from 71.3 to 73.7 tonnes/ha and 74.0 to 84.4 tonnes/ha, besides improving quality parameter, in Uttaranchal Uttar Pradesh, Rajasthan, Punjab, Haryana, and in 4 equal splits proved advantageous in the West Bengal. Tamil Nadu, Karnataka, 250 k- N/ha (cane yield 72.9 tonnes/ha) to 450 N/ha (cane \yield 123.3 tonnes/ha). The late shoots/ water shoots are normally harvested along with the millable canes. Retention of late shoots in crop harvested late (in April or May) provides additional benefit of cane yield in ratoon-crop. CoLK 8102 and CoLK 8001 registered 32.3 and 23.2 tonnes/ha higher cane yield when water- shoots/tillers were retained at the harvest.

Through photoperioic treatments, IND 85-581, IND 84-463, IND 84-461 and SES 293 clones were induced to flower. And similarly delay in flowering was also achieved. The delay in flowering based on first tip emergence ranged from 13 days in CoLK 8102 to 22 days in CoH 15 and BO 91. Clones of treated Saccharum sponataneum crosses flowered during second and third week of November, and control plants flowered in October.

Plant Protection. Co 7704, Co 7717, Co 802 1, CO 82 10, Co 860 10, Co 86011, Co 86250, Co 87267, Co 8842, Co 93009 and Co 93011 sugarcane varieties were found resistant red-rot. Biocontrol agents, Trichoderma sp. and Pseudomonas sp., were not able to control systematic infection of red-rot infected clumps. Saccharum robustum clones IM 76-258, NG 77-159, NG 87-213 and IS 76-137 and erianthus clones IS 76215, IJ 76-470, ERI 2798 and Timor Wild have showed multiple pest-resistance.

When the chopped bits of sugarcane having sufficient growth of red-rot fungus were incorporated in pots, and healthy sets were planted in red-rot debris incorporated soil it was observed that growth of red-rot fungus was more in the debris incorporated soil during early stage of planting.

In Punjab, release of Trichogramma chilonis at 50,000/ha at 10 days interval during May-June proved promising in planted and ratoon crops for controlling Chilo infuscatellus, and during July-October for Chilo auricilius. Topobracon sp. (1.7 2.8%) has been recorded for the first time on Scirpophaga exceptalis larvae in Punjab.

At Coimbatore, Sturmiopsis inferens peak activity on sugarcane shoot-borer chilo infuscatellus was noticed during January (5.6%) and November (0.1%) and of Cotesia flavipes was seen only in August (0.5%). Granulosis virus (GV) on shoot-borer larvae was seen throughout the year at Coimbatore and the peak was in August. Beauveria brongniartii showed limited ovicidal activity on white-grubs in laboratory. This fungus dosage equivalent of 1013 - 1017 spores/ha caused higher levels of mortality in the third instar grubs than first instars in pot experiments at Coimbatore.

Jute and Allied Fibres

Five strains of jute Corchorus capsularis have finer quality of fibre (1.30 Tex to 1.40 Tex) besides JRC 321 (1.55 Tex). Two strains of Corchorus olitorius are found at a par with Chaitali (JRO 878) in terms of fineness (2.97 Tex to 3.05 Tex).

Liming as well as farmyard manure (FYM) with recommended fertilizers produced highest C. capsularis fibre yield (1.83 tonnes/ha) when compared with tonnes/ha of control (N30 P30 9K30) at Nagaon (Assam). At the same location, in C. olitorius-rice cropping, 50 per cent inorganic nitrogen along with 5 tonnes of farmyard maure (FYM) /ha increased fibre yield (2.49 tonnes/ha) compared to 1.64 tonnes/ha in control.

Gliricidia (Gliricidia sepium) compost at 5 tonnes/ha increased fibre yield by 25 per cent over NO, P20, K20. And urea at 20 kg N/ha and Gliricidia compost at 2.5 tonnes/ha increased fibre yield by 42 per cent. Gliricidia compost enriched available soil N, and its residual effect on the succeeding crop was found promising.

Plant Protection. In jute, prevalence of root-rot was up to 30 per cent at Katihar (Bihar) and stem-rot and collar-rot up to 257(at Coochbihar (West Bengal). Inhibition of radial growth of Macrophomina phaseolina by Trichoderma harzianum (61%). Trichoderma viride (59%) and Gliocladium virens (44%) was observed in *in vitro.*

12

Farming in Cooperative Sector

The performance of crops vary considerably from location to location as well as from season to season. This is because of the influence of the environmental factors such as rainfall and temperature. It is difficult to control these environmental factors as we can control the factors like plant population, fertilizer rate and incidence of pests.

In order to determine the effects of uncontrollable environmental factors the experiments have to be repeated at different locations and seasons. With such repetitions of the experiments practical recommendations may be made with greater confidence, especially when new crop varieties or new technologies are introduced.

When the experiments are repeated at different locations the sample of location should be proper representative of the geographic area to which the results will apply. For this purpose the geographic area may be stratified according to agroclimatic conditions and soil factors, and the locations may be selected from these strata. In case of repetition of the experiments over seasons it is desirable that the seasons should be representative of the long term cycle of the seasons at a particular location.

Experiments in Agriculture

Once the locations or seasons are decided upon, the next step is to select the appropriate design of experiment. The individual experiments may be designed as randomised block, Latin square, split-plot or incomplete block designs. The same design is adopted for all locations or seasons. However, randomisation of treatments should be done afresh for each experiment.

Outcome Analysis

The results of the repeated experiments are analysed using combined analysis of variance method. The combined analysis is aimed at two objectives. One is to test whether there are significant differences between the treatments at various environments (locations, seasons, years, etc.).

The second is to test the consistency of the treatment responses at different environments, that is, to test the presence or absence of interaction of the treatments with environments. The absence of interaction between treatments and environments will indicate that the response to the treatment is consistent from environment to environment. The presence of interaction will indicate that the response changes with environment.

For the sake of convenience, only the location is taken as environment in the foregoing discussions. It can be substituted with seasons or years as the case may be.

Table: Combined analysis of variance based on RBD.

Sources of variation	*df*	*SS*	*MS*	*F*
Replications within location	p(r - 1)	R SS	R MS	
Location	p - 1	L SS	L MS	
Treatment	t - 1	T SS	T MS	TMS/LT MS
Location x Treatment	(p - 1) (t - 1)	LT SS	LT SS	LT MS/E MS
Pooled error	p(r - 1)(t - 1)	E SS	E MS	
Total	*rpt - 1*	*Total SS*		

In the first stage of the combined analysis the results of individual locations are analysed based on the basic experimental design used. In the second stage of the analysis various sums of squares are computed by combining all the data. The various steps involved in this process are explained by considering the basic design as randomised complete block design.

Before proceeding with the combined analysis, it is necessary to test whether the error mean squares of the individual experiments are homogeneous. The homogeneity of error mean squares can be tested by using either Barlett's test or Hartley's test. Hartley's F - maximum test is given by:

$$F - \text{max.} = \frac{\text{Larger error MS}}{\text{Smaller error MS}}$$

with (t - 1) and f degrees of freedom where t and f are the number of treatments and the error degrees of freedom in the experiment with higher mean square. As a working rule the error mean squares can be taken as homogeneous if F - max. is less than 3.

Unweighted Review

If the error mean squares are homogeneous the analysis is continued as follows:

$$CF = \frac{(\text{Grand total over all locations})^2}{\text{rpt}}$$

Total SS = (ΣY_{ij}^2 over all locations) - Cf

Replications within location SS = Sum of replication SS of all locations

Pooled error SS = sum of error SS of all locations.

The treatment x location two-way table is formed. From the two- way table treatment SS, location SS and treatment x location SS are computed as in the case of two factor experiments.

The significance of treatment x location interaction is tested by the criterion,

$$F = \frac{\text{treatment} \times \text{location MS}}{\text{Pooled error MS}}$$

If treatment x location interaction is significant the criterion

$$F = \frac{\text{treatment } MS}{T \times L \text{ interaction } MS}$$

is used to test the significance of the average effects of treatments. If the interaction is not significant the interaction *MS* and the pooled error *MS* are pooled to get a single denominator for the *F*-test. In practice, however, the first criterion is used even though the interaction is not significant. This is because of the reason that the interaction *MS* contains both the interaction and error components of variation which influence the treatment *MS*.

When the treatment x location interaction is significant, we may be interested in knowing why the performance of the treatments differed over locations. For this purpose, the treatment SS is partitioned into a set of orthogonal components like linear, quadratic, etc. The interaction SS is also partitioned in the same way. Then, the component of the treatment is tested against its own interaction with locations. When the error mean squares are homogeneous an alternative procedure is to analyse the data as split-plot design, taking treatments as the main plot treatments and the environments as subplot treatments.

Weighted Review

Weighted Analysis of Variance

Sources of variation	*df*	*MS*	*F*
Locations	$p - 1$	*LMS*	
Treatments	$t - 1$	*TMS*	*TMS/IMS*
L x T interaction	$(p - 1)(t - 1)$	*IMS*	
Total	$pt - 1$		

When the error mean squares are heterogeneous we resort to weighted analysis. In this case the structure of analysis of variance will be as shown in table above.

In weighted analysis the treatment x location table is formed first. The required sums of squares are then computed.

$$CF = \frac{(GT)^2}{t_w}$$

where, $GT = \Sigma(\Sigma W_i L_i)$

$$W_i = \left(\frac{r}{EMS}\right)_i = \text{weight for } i^{th} \text{ location}$$

L_i = total for i^{th} location

t - number of treatments

$$W = \Sigma W_i$$

Total SS = $\Sigma\{W_i \, \Sigma \, T_{ij}^2\} - CF$

where, T_{ij} = total for j^{th} treatment in i^{th} location

Location SS = $\frac{1}{t}\Sigma W_i (L_i^2) - CF$

Treatment SS = $\frac{1}{W}\Sigma(\Sigma W_i L_i)^2 - CF$

Treatment x Location SS = Total SS - location SS - Treatment SS

In case of heterogeneous error mean squares, the F-test is not valid to test the significance of the interaction between treatments and locations.

The valid test criterion is

$$\chi^2 = \frac{(k-4)(k-2)}{k(k+t-3)}.I$$

where, k = average error df

I = treatment x location MS

The degrees of freedom associated with this χ^2 is

$$(p-1).(t-1).\frac{(k-4)}{(k+t-3)}$$

where, p = number of locations or number of experiments combined.

If the interaction is proved to be present, the significance of the average effects of treatments is tested by the F-test where

$$F = \frac{\text{Treatment } MS}{\text{Interaction } MS}$$

The standard error for the comparison of treatment effects is given by

$$SE(d) = \sqrt{\frac{2(\text{interaction } MS)}{W}}$$

and, the critical difference is given by

$$CD = t \,.\, SE(d)$$

where, t = table value of t for interaction *d.f.* The treatment means are computed as

$$\bar{T} = \frac{1}{W}\Sigma W_i L_i$$

If the error mean squares are heterogeneous but the interaction is absent, the analysis of variance may be carried out as RBD using the average responses of the treatments. In such analysis the standard error for the comparison of treatment effects is given by

$$SE(d) = \sqrt{\frac{2(\text{interaction } MS)}{p}}$$

where, p = number of locations (or experiments pooled).

Farming in Cooperative Sector

Example : The data in table are from a suitability study on forage crops under dry farming conditions. The design adopted was RBD.

Greenmatter Yield, kg/plot

Variety	Year: 1982-83			
	Replication			
	I	II	III	IV
CN-1	2.85	2.55	4.55	3.54
CN-2	4.50	6.35	3.12	2.55
CN-3	3.15	5.45	2.85	2.15
BN-2	12.50	10.80	6.55	4.35
NB-21	10.60	8.95	5.80	4.22

Variety	Year: 1983-84			
	Replication			
	I	II	III	IV
CN-1	7.50	7.30	5.30	23.90
CN-2	4.25	14.35	6.80	14.00
CN-3	3.00	2.80	8.50	3.10
BN-2	21.00	5.70	2.80	9.40
NB-21	3.50	5.40	3.30	3.60

Variety	Year: 1984-85			
	Replication			
	I	II	III	IV
CN-1	8.60	10.30	18.40	8.10
CN-2	14.50	12.40	10.50	18.60
CN-3	4.40	5.80	7.20	10.00
NB-2	25.20	9.50	13.50	13.00
NB-21	9.50	9.10	20.40	11.60

The results of the RBD analysis for individual years are given in the following table.

Analysis of Variance for the Data in Table

Sources of variation	df	MS for the year		
		1982-83	1983-84	1984-85
Replication	3	14.3178	25.8795	18.1900
Treatment	4	23.6114	44.8700	42.2530
Error	12	3.4785	36.8820	24.3000

$$F-\max = \frac{36.8820}{3.4785} = 10.603$$

Since *F*-max is significant, it can be concluded that the error mean squares are heterogeneous. Hence, we proceed with weighted analysis. The required computations are given in the following table.

Computations in weighted analysis

Treatment	Year			$\Sigma W_i Y_i$
	Y_1	Y_2	Y_3	
T_1	13.49	44.00	45.40	27.757659
T_2	16.52	39.40	56.00	32.487887
T_3	13.60	17.40	27.40	22.036312
T_4	34.20	38.90	61.20	53.620230
T_5	29.57	15.80	50.60	44.045953
Year total: ΣY_i	107.38	155.50	240.60	179.948041
Rep. r_i	4	4	4	..
$(EMS)_i$	3.4785	36.8820	24.3000	..
$w_i = \left(\frac{r}{EMS}\right)$	1.149921	0.108454	0.164609	1.422984
$(\Sigma Y_i)^2$	11530.4644	24180.2500	57888.3600	93599.0744

ΣT_{ij}^2	2683.8754	5553.9700	12253.7200	20491.5654
$W_i(\Sigma Y_i)^2$	13.259.1232	2622.4448	9528.9451	25410.5131
$W_i \Sigma T_{ij}^2$	3086.2447	602.3503	2017.0726	5705.6676

In the above table,

$$\Sigma Y_1 = 13.49 + 16.52 + ... + 29.57 = 107.38$$

Similarly, $\Sigma Y_2 = 155.50$, and $\Sigma Y_3 = 240.60$

$$\Sigma T_{ij}^2 = (13.49)^2 + (16.52)^2 + ... + (29.57)^2 = 2683.8754$$

Similarly, $\Sigma T_{2j}^2 = 5553.9700$

$$\Sigma T_{3j}^2 = 12253.7200$$

For treatment 1,

$$\Sigma W_i Y_i = W_1 Y_1 + W_2 Y_2 + W_3 Y_3$$

$$= (1.149921)(13.49) + (0.108454)(44.00) + (0.164609)(45.40)$$

$$= 27.757659$$

Similarly, for other treatments we get, $\Sigma W_i Y_i$ values.

$$GT = \Sigma(\Sigma W_i Y_i) = 179.948041$$

$$CF = \frac{(G.T)^2}{tW} = \frac{(179.948041)^2}{5(1.422984)}$$

$$= 4551.1822$$

$$\text{Total } SS = \Sigma W_i\left(\Sigma T_{ij}^2\right) - CF$$

$$= 5705.6676 - CF$$

$$= 1154.4854$$

$$\text{Year } SS = \frac{1}{t}\Sigma W_i(\Sigma Y_i)^2 - CF$$

$$= \frac{1}{5}(25410.5131) - CF$$

$$= 530.9204$$

$$\text{Treatment } SS = \frac{1}{W}\Sigma\left(\left(\Sigma W_i Y_i\right)^2\right) - CF$$

$$= \frac{7126.7245}{1.422984} - CF$$

$$= 5008.2956 - CF$$

$$= 457.1134$$

Year x treat. $SS = I$ = Total SS - Year SS - Treat. SS

= 1154.4854-530.9204-457.1134 = 166.4516

k = average number of error degrees of freedom.

$$= \frac{(12+12+12)}{3} = 12$$

$$\chi^2 = \frac{(k-4)(k-2)}{k(k+t-3)}(I)$$

$$= \frac{(12-4)(12-2)}{12(12+5-3)}(166.4516)$$

$$= \frac{80}{168}(166.4516)$$

$$= 79.262$$

The df for $\chi^2 = (Y-1)(t-1)\dfrac{(k-4)}{(k+t-3)}$

$$= (3-1)(5-1)\frac{(12-4)}{(12+5-3)}$$

$$= 4.6$$

Weighted analysis of variance for the data in Table

Sources of variation	*df*	*SS*	*MS*	*F*
Year(r)	2	530.9204	265.4602	...
Treatment (T)	4	457.1134	114.2784	5.492
Y x *T*	8	166.4516	20.8064	
Total	14	1154.4854		

The table value of χ^2 has to be obtained by interpolation. In our example the χ^2 is highly significant. Hence the interaction is significant. The weighted analysis of variance as given in table has to be done in this case.

For weighted analysis the error degrees of freedom for individual experiments should be at least 15.

Then only the weights (W_i) will be sufficiently accurate. In our example this condition is not satisfied. However, for illustration purpose we have used the present data.

The standard error for the comparison of treatment effects is

$$SE(d) = \sqrt{\frac{2(IMS)}{W}}$$

$$= \sqrt{\frac{2(20.8064)}{1.422984}}$$

$$= \sqrt{29.24333653}$$

$$= 5.4077$$

$$CD = t\,.\,SE(d)$$

$$= (2.306)(5.4077) = 12.470$$

The approximate treatment means are given by $T_i = \frac{1}{W}\Sigma W_i Y_i$

For the present example,

T_1 = 19.51; T_2 = 22.83; T_3 = 15.49; T_4 = 37.68 and T_5 = 30.95.

Split-plot Design Series Experiment : So far we have discussed only the case of individual experiment in randomised blocks. The other type of design commonly used in agricultural experiments is spilt-plot design. The combined analysis for a series of split-plot experiments is described below.

The structure of combined analysis of variance for a series of split-plot experiments with factor A in main plots and factor *B* in subplots at *P* locations will be as in table below.

Table: Combined Analysis of Variance for Series of Split-plot Experiments

Sources of Variation	*df*	*MS*	*F*
Locations (*P*)	*p*-1	*P MS*	*P MS/R MS*
Replications within location	*p*(*r* - 1)	*R MS*	
Main plot treat. (*A*)	*a* - 1	*A MS*	*A MS/E MS*
P x *A*	(*p* - 1)(*a* - 1)	*PA MS*	*PA MS/E MS* (*a*)
Pooled error(a)	*p*(*r* - 1)(*a* - 1)	*E MS* (*a*)	
Subplot treat. (*B*)	6-1	*B MS*	*B MS/E MS* (*b*)
P x *B*	(*p* - 1) (*b* - 1)	*PB MS*	*PB MS/E MS* (*b*)
A x *B*	(*a* - 1) (*b* - 1)	*AB MS*	*AB MS/E MS* (*b*)
P x *A* x *B*	(*p* - 1) (*a* - 1) (*b* - 1)	*PAB MS*	*PAB MS/E MS* (*b*)
Pooled error (*b*)	*pa* (*r* - 1) (*b* - 1)	*E MS*(*b*)	
Total	*rpab* - 1		

In the first stage, the data from individual experiments are analysed. The homogeneity of both error *(a)* and error *(b)* are then tested. If the homogeneity of error *(a)* and error (*b*) are established, the combined analysis is carried out. When heterogeneity is present either in error *(a)* or error *(b)* or both the combined analysis is carried out.

Split-plot Experiment Combined Analysis when Error *(a)* and Error *(b)* are homogeneous

The required sums of squares are as follows:

Replications within locations *SS* = Sum of the replication sum of squares of individual experiments = $\sum RSS$

Pooled error (*a*) $SS = \sum ESS(a)$

Pooled error (*b*) $SS = \sum ESS(b)$

$$CF = \frac{(\text{Grand total over all locations})^2}{rpab}$$

$$\text{Total } SS = (\Sigma Y_{ij}^2 \text{ overall locations}) - CF$$

The main plot x location table, main plot x subplot table, location x subplot table, and location x main plot x subplot table are formed from which we compute the required *P SS*, *A SS*, *PA SS*, *B SS*, *AB SS*, *PB SS* and *PAB SS* in the usual way. The analysis of variance table is then completed.

Split-plot Experiments Combined Analysis when the Error Mean Squares are heterogeneous : If the homogeneity of error (a) mean squares is not established, the sums of squares involving factor *A* and the pooled error (a) *SS* are to be partitioned into linear, quadratic and so on, components using the orthogonal coefficients.

In case the error *(b)* mean squares are heterogeneous, the sums of squares involving factor *B* and pooled error *(b) SS* are partitioned into linear, quadratic, and so on, components.

The *F*-ratios are computed using the corresponding components.

Locations and Years Experiment Repetition : Agricultural field experiments are often repeated not only at different locations or years but both at a number of locations and for a number of years.

In such experimental programmes each year a new site is chosen at each selected location. Also a new randomisation is employed. This is done to ensure that the data from successive years are independent.

The structure of combined analysis of variance for a series of experiments in RBD involving *t* treatments, *p* locations and *y* years will be as given in the following table.

Table: Analysis of Variance for a Series of Experiments Conducted in RBD at a Number of Locations and for a Number of Years

Sources of Variation	*df*	*MS*
Locations (*P*)	p-1	*P MS*
Years (*Y*)	y-1	*Y MS*
$P \times Y$	$(p - 1)(y - 1)$	*PY MS*
Replications	$py(r - 1)$	*R MS*
Treatments (*T*)	t-1	*T MS*
$P \times T$	$(p - 1)(t - 1)$	*PT MS*
$Y \times T$	$(y - 1)(t - 1)$	*YT MS*
$P \times Y \times T$	$(p - 1)(y - 1)(t - 1)$	*PYT MS*
Pooled error	$py(t - 1)(r - 1)$	*E MS*
Total	$rpyt - 1$	

The test statistic to test the interaction between treatments, locations and years is,

$$F = \frac{PYT\,MS}{E\,MS}$$

The interaction between treatments and locations is tested by the test statistic,

$$F = \frac{PT\,MS}{PYT\,MS}$$

The interaction between treatments and years is tested by the test statistic,

$$F = \frac{YT\,MS}{PYT\,MS}$$

The test statistic for testing the consistency of treatment responses over locations is,

$$F = \frac{T\,MS}{PT\,MS}$$

The test statistic for testing the consistency of treatment responses over the years is

$$F = \frac{T\,MS}{YT\,MS}$$

If we are interested in testing the consistency of the treatments averaged over the locations and years the test statistic is

$$F = \frac{T\,MS}{E\,MS}$$

The above *F*-tests are appropriate only when the error mean squares and interaction mean squares of individual experiments are homogeneous. When these mean squares are heterogeneous, the analysis becomes more complicated. If the error mean squares are heterogeneous, the location x year x treatment interaction, may be tested by a weighted analysis of the means for each location and year, using weights as in the case in single series of experiments. When both location x treatment, and year x treatment interactions, are present the consistency of treatment responses may be tested using the criterion,

$$F = \frac{T\,MS + PYT\,MS}{PT\,MS + YT\,MS}$$

with

$$v_1 = \frac{(T\,MS + PYT\,MS)^2}{\dfrac{(T\,MS)^2}{t-1} + \dfrac{(PYT\,MS)^2}{(p-1)(y-1)(t-1)}}$$

and,

$$v_2 = \frac{(PT\,MS + YT\,MS)^2}{\dfrac{(PT\,MS)^2}{(p-1)(t-1)} + \dfrac{(YT\,MS)^2}{(y-1)(t-1)}}$$

degrees of freedom.

Bibliography

Agnew, J.A.: *Agriculture in India,* Allen & Unwin, New York, 1984.

Anderson, J.R.: *A Geography of Agriculture,* W.M.C Brown Co., Iowa, 1970,

Augelli, J.P.: *Middle America: Its Land and Peoples,* Prentice Hall, Englewood Cliff, 1989.

Bacon, R.S.: *The Cultural Landscape: An Introduction to Human Geography,* Prentice Hall, New Delhi, 1990.

Bansil, P.C.: *Agricultural Problems of India,* Vikas Publishing House, New Delhi, 1977.

Bhardwaj, N.: *Elements of Ecology,* Vikas Publications, New Delhi, 1979.

Binns, T.: *Tropical India,* Routledge, New York, 1994.

Biswas, M.R.: *Desertification: Environmental Science and Application,* Pergamon Press, New York, 1980.

Bose, A.: *India's Urbanization,* Kanishka Publication, New Delhi, 1980.

Boserup, E.: *The Conditions of Agricultural Crops,* Oxford, London, 1965.

Bourne, L.S.: *System of Indian Cities,* Oxford University Press, New York, 1978.

Brian, J.L.: *Geography of Market Centers and Retail Distribution,* Prentice-Hall, New Jersey, 1967.

Brian, P.: *Science in Geography,* Oxford University Press, London, 1974.

Briwn, J.: *India's Mineral Wealth,* Oxford University Press, New York, 1990.

Brown, R.: *Social Impact of the Green Revolution,* Oxford University Press, New York, 1971.

Buchanan, K.: *India: The Land and People,* Crown, New York, 1981.

Buck, J.L.: *Land Utilization in India,* Vikas Publication, New Delhi, 1967.

Burrard, S.G.: *A Sketch of the Geology and Geography of the Himalaya Mountains and Tibet,* Government of India, Delhi, 1933.

Cantwala: *Agricultural Development since Independence,* Oxford & IBH, New Delhi, 1966.

Carvalho, C.M.: *Readings in Cultural Geography,* Chicago University Press, Chicago, 1962.

Champion, H.G.: *Indian Forest Records,* Vikas Publication, New Delhi, 1936.

Chisholm, Michael: *Rural Settlement and Land Use: An Essay in Location,* Hutchinson, London, 1966.

Clapham, W.B.: *Natural Ecosystems,* MacMillan, London, 1973.

Clawson, D.L.: *India and Asia: Lands and Peoples,* Oxford, New Delhi, 2000.

Cole, J.: *Development and Underdevelopment: A Profile of the Third World,* Methuen, London, 1987.

Colin, J.: *Urban Geography: A First Approach,* Wiley & Sons, New York, 1982.

Coutinho, O.: *Economic and Commercial Geography of India,* Vikas Publishing House, New Delhi, 1988.

Cressey, G.B.: *Asia's Land and Peoples,* McGraw-Hill, New York, 1963.

Dadiani, N.K.: *The Hindu Survey of Indian Agriculture,* Kasturi & Sons Ltd., Chennai, 1999.

Dasseman, R.F.: *Environmental Conservation,* John Wiley, New York, 1976.

Demko, G.J.: *Population Geography: A Reader,* McGraw-Hill, New York, 1999.

Devaraj, M.: *The Hindu Survey of Indian Agriculture,* Kasturi & Sons, New Delhi, 1999.

Dey, A.K.: *Geology of India,* National Book Trust, New Delhi, 1968.

Dhillon, S.S.: *Agricultural Geography,* Tata McGraw-Hill Publication Co., Ltd. New Delhi, 1984.

Dikshit, K.R.: *India: Geomorphological Diversity,* Rawat Publications, Jaipur, 1994.

Dikshit, R.D.: *Geography and the Teaching of the Environment,* Department of Geography, Poona, University, 1984.

Dogra, Bharat: *Forest and People,* Himalaya Darshan Prakashan, Rishikesh, 1980.

Doshi, J.K.: *Bhils: Between Societal Self Awareness and Cultural Synthesis,* Sterling, New Delhi, 1971.

Doxiadis, C.A.: *Ekistics, An Introduction to the Science of Human Settlements,* Oxford University Press, New York, 1968.

Dubey, R.N.: *Economic Geography of India,* Kitab Mahal, Allahabad, 1998.

Duckham, A.N.: *Weather and Agriculture,* Oxford University Press, London, 1997.

Dumont, Rene: *Types of Rural Economy: Studies in Asia Agriculture,* Douglas Manin, Methuen, London, 1970.

Edgar, A.: *Plants, Man and Life,* University of California Press, California, 1967.

Edgar, S.: *Location of Agricultural Production,* University of Florida Press, 1954.

Ehrilich, P.R.: *Eco-Science,* W.H. Freeman & Co., San Franciscess, 1977.

Fade, Ole: *Geographic Perspectives on Migration Behaviour: A Bibliographical Survey,* MacMillan, London, 1979.

Farmer, B.H.: *Green Revolution,* MacMillan, London, 1977.

Febure, L.: *Geographical Introduction to History,* MacMillan, London, 1925.

Fellmann, J.: *Human Geography,* MacMillan, New York, 1989.

Found, W.C.: *Theoretical Approach to Rural Land Use Patterns,* St. Martin, New York, 1971.

Frankel, F.R.: *India's Green Revolution: Agriculture Costs of Land Growth,* Princeton University Press, Princeton, 1971.

————————: *India's Green Revolution: Political Costs of Economic Growth,* Princeton University Press, Princeton, 1971.

Gandhi, M.P.: *Major Industries of India, 1970-71,* M.P. Gandhi and Co., Bombay, 1972.

Gansser, A.: *Geology of the Himalayas,* John Wiley & Sons, London, 1964.

Gill, K.S.: *Evolution of the Indian Agriculture,* N.C.E.R.T., New Delhi, 1993.

Gopalakrishnan, Ramamoorthy: *Geography of India,* Jawahar Publishers & Distributors, New Delhi, 1996.

Gorsoline, D.: *North American Indians,* Random House, New York, 1978.

Goudie, A.: *The Human Impact on the Natural Environment,* Basil Blackwell, Oxford, 1986.

Gregory, D.: *Ideology, Science and Human Geography,* Hutchinson, London, 1978.

————————: *The Geographical Imagination, Social Theory and Human Geography,* Hutchinson, London, 1985.

Griffin, Keith B.: *The Political Economy of Agrarian Change: An Essay on the Green Revolution,* Harvard University Press, Harvared, 1974.

Grigg, D.B.: *Geographical Impact of Migration,* White and Woods, Longman, London, 1980.

————————: *The Agricultural Systems of the India,* Cambridge University Press, 1973.

Gyorgy, Enyedi: *Geographical Types of Agriculture,* Applied Geography in Hungary, Budapest, 1964.

Haddon, A.C.: *The Races of Man and their Distribution,* Oxford, New York, 1925.

Haggett, P.: *Geography:A Modern Synthesis,* Harper and Row, London, 1972.

————————: *Socio-Economic Models in Geography,* Methuen, London, 1967.

Hartshorn, T.A.: *Interpreting the City: An Urban Geography,* John Wiley & Sons, New York, 1990.

Hartshorn, T.A. and Alexander, J.W.: *Economic Geography,* Prentice Hall, London, 1988.

Harvey, D.W.: *Explanation in Geography,* Arnold, London, 1969.

Hathaway, D.E.: *Government and Agriculture: Public Policy in a Democratic Society,* MacMillan, London, 1963.

Haviland, M.D.: *Forest, Steppe and Tundra,* Cambridge, London, 1926.

Heathcote, R.L.: *The Arid Lands: Their Use and Abuse,* Longman, New York, 1983.

Henshall, J.D.: *Models in Geography,* Methuen, London, 1967.

Hoover, E.M.: *Location of Economic Activity,* McGraw Hill, New York, 1948.

Hopf, M.: *Agricultural Geography,* Clarendon Press, Oxford, 1988.

Hoskins, W.G.: *The Making of English Landscape,* Hodder and Stoughton, London, 1988.

Hourani, G.F.: *Arab Seafaring in the Indian Ocean and Early Medieval Times,* Princeton University Press, Princeton, 1951.

Howard, F.: *Geography of Agriculture: Themes in Research*, Prentice-Hall, New Jersey, 1970.

————: *Geography of Agriculture: Themes in Research*, Prentice-Hall, New Jersey, 1970.

Hoyle, S.B.: *Spatial Aspects of Development*, John Wiley, New York, 1974.

Husain, M.: *Evolution of Geographical Thought*, Rawat, Jaipur, 1995.

————: *Human and Economic Geography*, NCERT, New Delhi, 1928.

Ismail, R.: *Historical Atlas of the Religions of the World*, MacMillan, New York, 1974.

Jacob, J.: *The Economy of Cities*, Vintage Books, New York, 1970.

James, A.: *Weather and Agriculture*, Pergamon, Oxford, 1967.

James, L.: *Resource Geography*, Prentice Hall, New Jersey 1979.

James, O. and Muller, P.O.: *Economic Development*, Wiley, New York, 1981.

Jenkins, M.M.: *The Himalayan Kingdoms: Bhutan, Sikkim and Nepal*, Van Mostrand Company, New Jersey, 1963.

Jhingran, A.G.: *Geology of the Himalayas*, Oxford University Press, New Delhi, 1981.

Johnson, B.L.: *India: Resources and Development*, Arnold - Heinemann, London, 1980.

Johnston, R.J.: *Geography and Geographers*, Oxford University Press, New York, 2000.

————: *A Dictionary of Human Geography*, Black-well, Oxford, 1990.

Jone, H.R.: *A Population Geography*, Harper and Row, London, 1981.

Jones, E.: *Human Geography*, Chatto & Windus, London, 1972.

Jordan, T.G.: *The Indian Agriculture Area: A Resource Geography*, Rawat Publication, Jaipur, 1996.

Joshi, H.L.: *Agricultural Geography of India,* Rawat Publications, Jaipur, 1990.

Joshi, J.K.: *Social Structure and Cultural Change in a Bhil Village,* Heights, New Delhi, 1974.

Krishan, G.: *Geography of India,* Rawat Publications, Jaipur, 1996.

Krishnan, M.S.: *Geology of India and Burma,* Rima Publishing House, Delhi, 1982.

Kulkarni, M.R.: *Industrial Development,* National Book Trust, New Delhi, 1991.

Lal, D.S.: *Climatology,* Chaitanya Publishing House, Allahabad, 1986.

Laxmi, Abha: *The Problem of Wastelands in India,* B.R. Publishing Corporation, Delhi, 1985.

Learmonth, A.T.A.: *India and Pakistan: A General and Regional Geography,* Methuen & Co. Ltd., London, 1967.

————————: *India and Pakistan: A General and Regional Geography,* Methuen, London, 1971.

Lewis, G.J.: *Human Migration,* MacMillan, London, 1982.

Longrigg, S.H.: *Resource Geography of India,* Rawat Publication, Jaipur, 1970.

Lyman, H.: *Beginnings of Agriculture in India,* Vikas Publication, New Delhi, 1968.

Major, John, S.: *The Land and People of India,* Lippincott, Philadelphia, 1989.

Majumdar, D.N.: *Races and Cultures of India,* Asia Publishing House, Bombay, 1958.

Manivasakam, N.: *Environmental Pollution,* N.B.T., New Delhi, 1992.

Manmion, A.M.: *Global Environmental Change,* Longman, New York, 1991.

Mansfield, G.B.: *Farming Systems of the World,* Praeger, New York, 1970.

Martin, G.J.: *All Possible Worlds: A History of Geographical Ideas,* John Wiley & Sons, New York, 1993.

Maxwell, K.E.: *Environment of Life,* Dickenson Publishers, Co., California, 1973.

McRae, S.C. and Burnham, C.P.: *Land Evaluation,* Oxford University Press, New York, 1981.

Mehar, D.N.: *Minerals of India,* National Book Trust, New Delhi, 1994

Mehta, J.K.: *Present State of Indian Economy,* Oriental Publishers, Allahabad, 1977.

Miracle, M.P.: *Traditional Agricultural Methods in India,* Stanford: Food Research Institute, New York, 1964.

Misra, R.P.: *Development Issues of our Time,* Concept, New Delhi, 1984.

Misra, V.C: *India-Regional Studies,* 21st IGU Publication, New Delhi, 1968.

Morgan, G.C.: *Human and Economic Geography,* Oxford University Press, Oxford, 1992.

Mumford, L.: *The City in History,* Harcourt, Brace & World, New York, 1961.

Munton, R.J.: *Agricultural Geography,* Methuen & Co. London, 1972.

Muthiah, S.: *An Atlas of India,* Oxford University Press, New York, 1992.

Myrdal, Gunnar: *Rich Lands and Poor,* Harper & Brothers, New York, 1957.

Nag, P. and Sengupta, S.: *Geography of India,* Concept Publication Co., New Delhi, 1992.

Narayan, L.: *Wastelands in India,* Pink, Mathura, 1988.

Newman, J.A.: *Population Patterns: Dynamics and Prospects,* Prentice Hall, New Jersey, 1984.

Ojha, S.S.: *Geography of India,* Bhaugolik Adhyayan Sansthan, Allahabad, 2001.

Oliver, D.L. *Native Cultures of the Pacific Islands,* University of Hawai Press, Honolulu, 1989.

Paliwal, Kamlesh: *Economic Geography of India,* Kitab Mahal, Allahabad, 2000.

Pannikkar, K.M.: *India and the Indian Ocean: An Essay on the Influence of Sea Power on Indian History,* George Allen and Unwin, London, 1951.

Parain, C.: *The Evolution of Agricultural Technique* Cambridge University Press, Cambridge, 1966.

Parakh, B.S.: *India: Economic Geography,* N.C.E.R.T., New Delhi, 1990.

Parmer, B.H.: *The Green Revolution in India,* The MacMillan, London, 1980.

Paul, A.: *Introduction to Agriculture,* John Wiley and Sons, New York, 1973.

Paul, Knox: *Urban Social Geography,* Longman, New York, 1987.

Peach, C.: *Geography and Ethnic Pluralism,* Allen & Unwin, London, 1984.

Pedelaborde, P.: *The Monsoon,* Methuen and Co., London, 1963.

Persson, G.A.: *Resources and World Development,* John Wiley, Chichester, 1987.

Ponekar, G.S.: *Studies in Green Revolution,* United Asia Publications, Bombay, 1970.

Pritchard, J.M.: *Landform and Landscape in India,* Edward Arnold, London, 1979.

Puri, G.S.: *Indian Forest Ecology,* Oxford, New Delhi, 1960.

Pyke, M: *Man and Food,* McGraw Hill, New York, 1970.

Qureshi, M.H.: *India: Resources and Regional Development,* N.C.E.R.T., New Delhi, 1990.

Rabb, T.: *Hunger and History: The Impact of Changing Food Production and Consumption Patterns on Society*, Cambridge University Press, New York, 1984.

Rahdhawa, M.S.: *Agriculture and Animal Husbandry in India*, ICAR, New Delhi, 1958.

Rao, B.P.: *Bharat: Ek Bhaugolik Samiksha*, Vasundhara Prakashan, Gorakhpur, 1995.

Rao, K.L.: *India's Water Wealth*, Orient Longman, New Delhi, 1975.

Raychaudhuri, S.P.: *Land and Soil*, National Book Trust of India, New Delhi, 1966.

Richard, J.: *Network Analysis in Geography*, Edward Arnold, London, 1969.

Sauer, C.O.: *Agriculture Origins and Dispersals*, American Geographical Society, New York, 1952.

Sen, Bandhudas: *The Green Revolution in India: A Perspective*, Wiley Eastern, New Delhi, 1974.

Sen, L.K.: *Planning Rural Growth Centres for Integrated Area Development: A Study in Miryalguda Taluka*, N.I.C.D., Hyderabad, 1971.

Shafi, M.: *Agricultural Productivity and Regional Imbalances*, Concept Publication Co. New Delhi, 1984.

———————: *Ganga-Yamuna Doab, in India: Regional Studies*, National Committee for Geography, Calcutta, 1923.

Shiva, V.: *The Violence of Green Cultivation*, Mapusa, Goa, 1991.

Siddhartha, K.: *India: the Physical Aspects*, Centre for Development of Environment and Resources, Patna, 1999.

Singh, J.: *Agricultural Geography*, Tata McGraw Hill Publication Co., New Delhi, 1980.

———————: *An Agricultural Atlas of India: A Geographical Analysis* Vishal Publication, Kurukshetra, 1974.

———————: *An Agricultural Atlas of India: A Geographical Analysis*, Vishal Publications, Kurukshetra, 1974.

Singh, R.L.: *India: A Regional Geography,* National Geographical Society of India, Varanasi, 1971.

Singh, Surendra: *Agricultural Development in India: A Regional Analysis,* Kaushal Publications, Shillong, 1994.

——————: *Agricultural Development in India: A Regional Analysis,* Kaushal Publications, Shillong, 1994.

Small, R.J.: *The Study of Landforms,* Cambridge University Press, London, 1970.

Smith, C.J.: *China: People and Places in the Land of India,* Rawat Publication, Jaipur, 1984.

Smith, S.J.: *Exploring Social Geography,* Allen & Unwin, London, 1984.

Srinivasan. T.N.: *The Green Revolution,* Indian Statistical Institute, Calcutta, 1971.

Stamp, L.D.: *The Land of India: its Use and Misuse,* Vikas Publication, New Delhi, 1948.

Sundaram, K.V.: *Planning Revisited: Rural Area Development in India: Vision and Reality,* R.N. Dubey Memorial Lectures, Allahabad, 1999.

Symon, L.: *Agricultural Geography,* London, 1968.

Taylor, G.: *Geography in the 20th Century,* Methuen, London, 1967.

Taylor, P.J.: *New Horizons in Geography,* MacMillan, London, 1986.

Terra, H.: *Studies on the Ice Age in India,* Washington, 1995.

Theakstone, W.H.: *The Analysis of Geographical Data,* Heinemann, London, 1970.

Thomas, C.: *A History of the Land and Agriculture,* Prentice Hall, Englewood Cliff, 1978.

Thomas, R.: *Man's Impact on the Environment,* McGraw-Hill, New York, 1988.

Thomas, W.L.: *Man's Role in Changing the Face of the Earth,* University of Chicago Press, 1965.

Tiwari, R.T.: *Development and Change is India,* Ashish Publication, New Delhi, 1968.

Unwin, Tim,: *Indian Geography,* Harlow, New York, 1998.

Van, F.: *Indian Foreign Policy and the Border Dispute,* The Hague, London, 1967.

Wadia, D.N.: *Geology of India,* Tata McGraw-Hill Publishing Co., New Delhi, 1975.

Wadia, F.K.: *Agro-Climatic Regional Planning in India,* Concept Publishing Company, New Delhi, 1996.

Wegener, A.: *The Origin of Continents and Oceans,* John Biram, New York, 1966.

Wrighly, E.A.: *Population and History,* Oxford University Press, New York, 1969.

Zelinsky, W.: *A Handbook of Agriculture,* Prentice Hall, New Jersey, 1966.

Zhao, S.: *Geography of India,* John Wiley & Sons, New York, 1986.

Zudaic, P.: *India: A Geographical Study,* John Wiley & Sons, New York, 1988.

Index

F

M

N

O

P

❑❑❑